MW00680742

Student Solutions

Engineering Statistics
The Industrial Experience

Bernard Ostle
University of Central Florida

Kenneth V. Turner
Anderson University

Charles Hicks
Purdue University

Gayle W. McElrath
President, McElrath and Associates, Inc.

Duxbury Press

An Imprint of Wadsworth Publishing Company
I(T)P® An International Thomson Publishing Company

Belmont, CA • Albany, NY • Bonn • Boston • Cincinnati • Detroit • Johannesburg • London
Madrid • Melbourne • Mexico City • New York • Paris • San Francisco • Singapore • Tokyo
Toronto • Washington

International Thomson Publishing Europe
Berkshire House 168-173
High Holborn
London, WC1V 7AA, England

International Thomson Editores
Campos Eliseos 385, Piso 7
Col. Polanco
11560 México D.F. México

Thomas Nelson Australia
102 Dodds Street
South Melbourne 3205
Victoria, Australia

International Thomson Publishing Asia
221 Henderson Road
#05-10 Henderson Building
Singapore 0315

Nelson Canada
1120 Birchmount Road
Scarborough, Ontario
Canada M1K 5G4

International Thomson Publishing Japan
Hirakawacho Kyowa Building, 3F
2-2-1 Hirakawacho
Chiyoda-ku, Tokyo 102, Japan

International Thomson Publishing GmbH
Königswinterer Strasse 418
53227 Bonn, Germany

International Thomson Publishing Southern
Africa
Building 18, Constantia Park
240 Old Pretoria Road
Halfway House, 1685 South Africa

ISBN 0-534-26539-1

CONTENTS

CHAPTER 1
STATISTICS: A TOOL FOR DECISION MAKING

Section 1.1

1.1 (a) The universe is the set of cola drinkers in the vicinity of Anderson, Indiana. The experimental units are the individual cola drinkers.

(b) Letting N denote the number of persons in the universe, the population of interest is a set of K yeses and $N - K$ noes, with K the number who prefer Coca-Cola®. The parameter of interest is $\theta = K/N$, the true proportion of cola drinkers who prefer Coca-Cola®.

(c) The sample of experimental units is the set of persons who participate in the taste test. Letting n denote the number participating in the taste test, the sample of the population is a set of x yeses and $n - x$ noes, with x the number who prefer Coca-Cola. A sample statistic is $p = x/n$, the proportion of those participating who prefer Coca-Cola.

(d) The experimenter is likely to want to use the value of p to infer that the true proportion of cola drinkers in the vicinity of Anderson, Indiana who prefer Coca-Cola to the other cola used in the study is greater than 0.50.

1.3 (a) The null hypothesis is H_0: "The person is telling the truth." The alternative hypothesis is H_a: "The person is not telling the truth."

(b) A Type I error occurs when the polygraph indicates a truth-sayer is not telling the truth.

(c) It appears that a Type I error occurs in about $\left(\frac{185}{500} \times 100\right)\% = 37\%$ of the tests involving a person who is telling the truth.

(d) A Type II error occurs when a person lies and the polygraph does not indicate that a lie has occurred.

(e) It appears that a Type II error occurs in about $\left(\frac{120}{500} \times 100\right)\% = 24\%$ of the tests involving a person who is not telling the truth.

(f) The seriousness of the error depends on your point of view. A person who is called a liar and sentenced to jail because a Type I error occurs would consider the error to be quite serious, as should the court. However, a liar who is not sentenced because a Type II error occurs is, unlike the court, not likely to be concerned about that error.

1.5 Let L denote the list of telephone numbers being used by the callers. Universes include: the houses, apartments, buildings, and so forth associated with the telephone numbers in L; the persons in the households associated with telephone numbers in L; the heads of the households associated with telephone numbers in L; and the people who are home between 12:00 and 4:00 p.m. on the Saturday the calls are made and will answer the phone when it rings.

Sections 1.3.1 and 1.3.2

1.13 From the process analysis map on page 3, it appears that deferred
maintenance has been practiced. Most of the causes cited are related to
the facilities and the equipment. As soon as they come available,
resources should be invested to repair (or replace) faulty equipment.
Estimates of the losses associated with the different causes could be used
to aid in determining the first steps to be taken. If the equipment is in
disrepair to the point that material is damaged due to lack of control by
the operators, a safety hazard may exist. It would be cheaper to repair
or replace faulty equipment than to engage in a lawsuit over safety
violations. A foreman or supervisor should be assigned the task of
making certain the crane operators are properly trained.

Section 1.3.3

1.15 The contour problem accounts for 79% of the scrap loss. The primary
emphasis should be on correction of that problem.

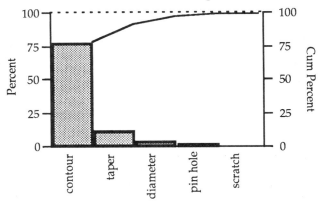

1.17 For the 24 hoods in the study, the major problem is dirt. It accounts for
over 50% of the incidences recorded.

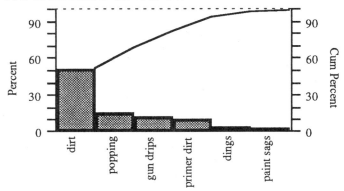

Problem	Level I causes	Level II causes	Level III causes

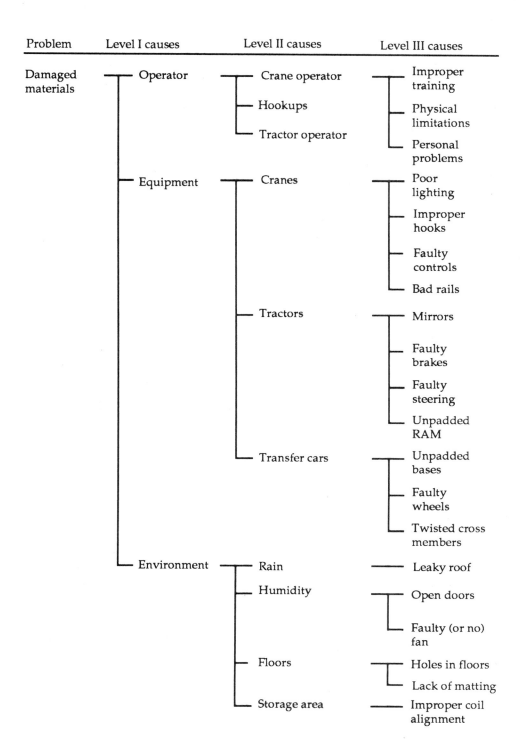

1.19   The 48 channels meet the specifications of 50 to 120 millivolts. The variability within controller 1 is much greater than that of the other controllers. The reading of 108.8 for the left front (LF) channel of that controller is the largest of all the readings. A check should be made to see if a recording error was made, or if some unusual condition existed for controller 1.

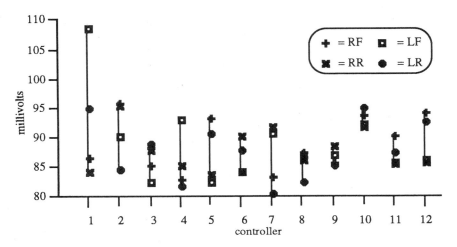

1.21   Since the taste-test was performed to estimate the proportion of cola drinkers in the population of all cola drinkers in the vicinity of Anderson, Indiana who prefer coke, the study is enumerative.

1.25   The Level I cause given as *quality costs* may indicate that many are not committed to the same goals. Deming's first point addresses the need for *constancy of purpose*.

The listing of quality costs as a Level I cause may indicate that many salaried employees do not realize the advantages in providing product of higher quality. This is addressed in Deming's second point.

It appears that an attempt is being made to provide quality products by adding quality on, rather than building quality in. Dependence on mass inspection needs to be addressed (see point 3).

The adverse relations, mistrust, inequities, and lack of communication listed in the process analysis map are addressed in several of Deming's fourteen points. Instituting modern methods of training and supervision, driving out fear throughout the organization, breaking down barriers among departments, eliminating goals that cannot be achieved, and removing barriers that prevent people from taking pride in their jobs address those issues.

4

CHAPTER 2

DESCRIPTIVE STATISTICS AND GRAPHICAL DISPLAYS

Section 2.1

2.1   **(a)**

4t	333
4f	4444555
4s	666666666666666677777777
4•	888888888888888889999999999999999
5*	0000000000000001
5t	33

outside value = 97

     (b)   On the diagram in (a), we noted that 97 is an outside value. The reason for its being such should be investigated, and further analysis should continue with that value removed.

2.3

37*	56679
38•	00112333
38*	777789
39•	0000001112233344
39*	55555555777778
40•	0000133
40*	556

outside = 491

A stem-and-leaf display with double stems is shown. The data seem to have been obtained from a reasonably symmetric distribution. The outside value, 491, is likely due to a recording error— 391 or 401 seem to be more likely values. The reason for 491 being included in the data should be investigated.

Section 2.2

2.5   The engineer who set up the simulation obtained the displayed histogram. After seeing the large number of peaks and valleys, he questioned the validity of the simulation. Further investigation revealed that loop times for many functions, rather than a single function, had been obtained.

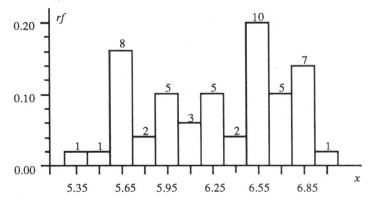

5

2.7   **(a)**   The range is $R = x_{max} - x_{min} = 5.5 - 3.0 = 2.5$. Beginning at 2.75, and using an interval width of 0.50, gives the 6 class intervals in the accompanying frequency distribution.

class	tally	f	rf	cf	rcf
[2.75, 3.25]	I	1	0.01	1	0.01
[3.25, 3.75]	II	2	0.02	3	0.03
[3.75, 4.25]	IIIII IIIII IIIII IIIII IIIII IIIII IIIII III	38	0.38	41	0.41
[4.25, 4.75]	IIIII IIIII IIIII IIIII IIIII IIIII IIIII IIIII	40	0.40	81	0.81
[4.75, 5.25]	IIIII IIIII III	13	0.13	94	0.94
[5.25, 5.75]	IIIII I	6	0.06	100	1.00

(b)   The following relative frequency histogram indicates that the sampled distribution is reasonably symmetric, centered near 4.25 pounds. Since all measurements are below 5.75, it appears that almost no lever efforts associated with the sampled air controllers exceed 6.0 pounds.

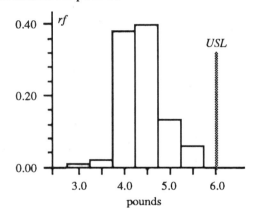

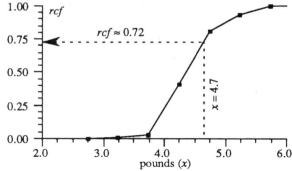

(c)   Using the preceding ogive, the lever efforts of about 72% of the sampled air controllers are 4.7 pounds or less.

2.9 **(a)** The range is $R = x_{max} - x_{min} = 9.6 - 0.7 = 8.9$. Beginning at 0.15, and using an interval width of 1.10, gives the 9 class intervals in the following frequency distribution.

class	tally	f	rf	cf	rcf
[0.15, 1.25]	IIII IIII I	11	0.088	11	0.088
[1.25, 2.35]	IIII IIII IIII IIII IIII I	26	0.208	37	0.296
[2.35, 3.45]	IIII IIII IIII IIII IIII IIII I	31	0.248	68	0.544
[3.45, 4.55]	IIII IIII IIII IIII IIII IIII I	31	0.248	99	0.792
[4.55, 5.65]	IIII IIII IIII II	17	0.136	116	0.928
[5.65, 6.75]	IIII	5	0.040	121	0.968
[6.75, 7.85]	II	2	0.016	123	0.984
[7.85, 8.95]	I	1	0.008	124	0.992
[8.95, 10.05]	I	1	0.008	125	1.000

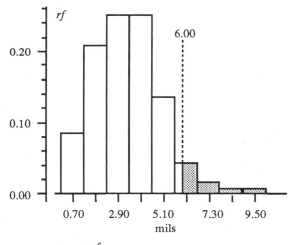

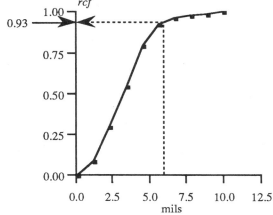

(b) The preceding relative frequency histogram indicates that the sampled distribution is skewed to the right. The shaded region indicates that the upper specification limit should not be decreased.

(c) Using the preceding ogive, about 7% of the sampled circuit boards have cleanliness readings that exceed 6.0 mils.

Section 2.3

2.11 **(a)** $\bar{x} = (7.5 + 3.0 + 9.6 + 1.6 + 1.8)/5 = 23.5/5 = 4.7$

(b) The ordered values are $x_{(1)} = 1.6$, $x_{(2)} = 1.8$, $x_{(3)} = 3.0$, $x_{(4)} = 7.5$, and $x_{(5)} = 9.6$. Since $(0.25)(5 + 1) = 1.5$, $(0.5)(5 + 1) = 3$, and $(0.75)(5 + 1) = 4.5$, the sample quantiles are:
$q_1 = x_{0.25} = x_{(1)} + (0.5)(x_{(2)} - x_{(1)}) = 1.6 + (0.5)(1.8 - 1.6) = 1.7$,
$\tilde{x} = q_2 = x_{0.50} = x_{(3)} = 3.0$, and
$q_3 = x_{0.75} = x_{(4)} + (0.5)(x_{(5)} - x_{(4)}) = 7.5 + (0.5)(9.6 - 7.5) = 8.55$.

(c) Since $(0.80)(5 + 1) = 4.80$, the 80th percentile is
$x_{0.80} = x_{(4)} + (0.80)(x_{(5)} - x_{(4)}) = 7.5 + (0.80)(9.6 - 7.5) = 9.18$.

(d) Since $100[2/(5 + 1)] \approx 33.3$, $x_{(2)} \approx x_{0.333}$.

2.13 Since $\bar{x}_{new} = [(7.2)(5) + 8 + 7]/7 = 51/7 = 255/35$ and $\bar{x}_{old} = 7.2 = 252/35$, the sample average increased $3/35$ inch-pound.

2.15 The information used to answer (b), (c), and (d) was obtained by using a computer to order the data in prb02_06 on the data disk.

(a) $\bar{x} = \sum x_i/64 = 3673/64 \approx 57.4$

(b) Since $(0.25)(64 + 1) = 16.25$, $(0.50)(64 + 1) = 32.50$, and $(0.75)(64 + 1) = 48.75$, the sample quartiles are:
$q_1 = x_{0.25} = x_{(16)} + (0.25)(x_{(17)} - x_{(16)}) = 55 + (0.25)(55 - 55) = 55$
$\tilde{x} = q_2 = x_{0.50} = x_{(32)} + (0.50)(x_{(33)} - x_{(32)}) = 57 + (0.50)(57 - 57) = 57$
$q_3 = x_{0.75} = x_{(48)} + (0.75)(x_{(49)} - x_{(48)}) = 60 + (0.75)(60 - 60) = 60$

(c) Since $(0.65)(64 + 1) = 42.25$, the 65th percentile is
$x_{0.65} = x_{(42)} + (0.25)(x_{(43)} - x_{(42)}) = 58 + (0.25)(59 - 58) = 58.25$.

(d) Since $100[23/(64 + 1)] \approx 35.4$, $x_{(23)} \approx x_{0.354}$.

2.17 The information used to answer (b), (c), and (d) was obtained by using a computer to order the data in prb02_08 on the data disk.

(a) $\bar{x} = \sum x_i/110 = 495.8/110 \approx 4.51$

(b) Since $(0.25)(110 + 1) = 27.75$, $(0.50)(110 + 1) = 55.50$, and $(0.75)(110 + 1) = 83.25$, the sample quartiles are:
$q_1 = x_{0.25} = x_{(27)} + (0.75)(x_{(28)} - x_{(27)}) = 4.0 + (0.75)(4.0 - 4.0) = 4.0$
$\tilde{x} = x_{0.50} = x_{(55)} + (0.50)(x_{(56)} - x_{(55)}) = 4.5 + (0.50)(4.5 - 4.5) = 4.5$
$q_3 = x_{0.75} = x_{(83)} + (0.25)(x_{(84)} - x_{(83)}) = 5.0 + (0.25)(5.1 - 5.0) \approx 5.03$

(c) Since $(0.44)(110 + 1) = 48.84$, the 44th percentile is
$x_{0.44} = x_{(48)} + (0.84)(x_{(49)} - x_{(48)}) = 4.5 + (0.84)(4.5 - 4.5) = 4.5$.

(d) Since $100[68/(110 + 1)] \approx 61.3$, $x_{(68)} \approx x_{0.613}$.

Section 2.4

2.19 $\bar{x} = \Sigma x_i/6 = (3.8 + 4.2 + 4.3 + 4.0 + 5.2 + 4.3)/6 = 25.8/6 = 4.3$
The ordered values are: $x_{(1)} = 3.8$, $x_{(2)} = 4.0$, $x_{(3)} = 4.2$, $x_{(4)} = 4.3$, $x_{(5)} = 4.3$, and $x_{(6)} = 5.2$.
$\tilde{x} = [x_{(3)} + x_{(4)}]/2 = [4.2 + 4.3]/2 = 4.25$
$R = x_{max} - x_{min} = 5.2 - 3.8 = 1.4$
$$SSX = (3.8 - 4.3)^2 + (4.2 - 4.3)^2 + (4.3 - 4.3)^2 + (4.0 - 4.3)^2 + (5.2 - 4.3)^2$$
$$+ (4.3 - 4.3)^2$$
$$= (-0.5)^2 + (-0.1)^2 + 0 + (-0.3)^2 + (0.9)^2 + 0 = 1.16 \text{ [Equation (2.5)]}$$
$s^2 = SSX/(n-1) = 1.16/5 = 0.232$ and $s = \sqrt{0.232} \approx 0.48$

2.21 $\bar{x} = \Sigma x_i/5 = (13 + 10 + 10 + 5 + 13)/5 = 51/5 = 10.2$
The ordered values are: $x_{(1)} = 5$, $x_{(2)} = 10$, $x_{(3)} = 10$ $x_{(4)} = 13$, and $x_{(5)} = 13$.
$\tilde{x} = x_{(3)} = 10$ and $R = x_{max} - x_{min} = 13 - 5 = 8$
$$s^2 = [(13 - 10.2)^2 + (10 - 10.2)^2 + (10 - 10.2)^2 + (5 - 10.2)^2 + (13 - 10.2)^2]/4$$
$$= [(2.8)^2 + (-0.2)^2 + (-0.2)^2 + (-5.2)^2 + (2.8)^2]/4 = 42.8/4 = 10.7$$
$s = \sqrt{10.7} \approx 3.3$

2.23 Using a computer and the data in prb02_22, $q_3 = 31$, $q_1 = 28$, $IQR = 31 - 28 = 3$, and $x_{0.90} = 31$.

2.25 From problem 2.22, $\bar{x} \approx 29.2$ and $s \approx 1.6$. Thus, the standardized value of 31 is $z \approx (31 - 29.2)/1.6 \approx 1.1$. This indicates that 31 is about 1.1 standard deviations above the mean.

2.27 Using a computer to summarize and order the data in prb02_26, we find $q_3 = 55$, $q_1 = 53$, $IQR = 55 - 53 = 2$, $x_{(21)} = 55$, and $x_{(22)} = 55$. Since $(0.84)(26) = 21.84$, the 84th percentile is $x_{0.84} = x_{(21)} + (0.84)(x_{(22)} - x_{(21)}) = 55$.

2.29 From problem 2.26, $\bar{x} \approx 53.9$ and $s \approx 1.8$. Thus, the standardized value of 56 is $z \approx (56 - 53.9)/1.8 \approx 1.2$. This indicates that 56 is about 1.2 standard deviations above the mean.

2.31 $\bar{x} = [42(3) + 45(22) + 48(43) + 51(16) + 54(2)]/[3 + 22 + 43 + 16 + 2] = 4,104/86 \approx 47.7$, $\tilde{x} = 48$ and $R = 54 - 42 = 12$.

Since $SSX = [(42)^2(3) + (45)^2(22) + (48)^2(43) + (51)^2(16) + (54)^2(2)]$
$$- [(4,104)^2/86] \approx 515.3023,$$
$s^2 = SSX/85 = 515.3023/85 \approx 6.0624$ and $s = \sqrt{6.0624} \approx 2.5$.

2.33 **(a)** If each value is increased by 10 units, the mean of the new values will be 110. The variance of the new values will be the same as that for the original data, 15.
   (b) If each value is multiplied by 10 units, the mean of the new values will be 10(100) = 1000. The variance of the new values will be (100)(15) = 1500.

2.35 Using Chebyshev's rule, at least 75% of the values of any distribution are within 2 standard deviations of the mean. Since we know nothing specific about the distribution (bell-shaped, one mode, and so forth), the correct choice is (e).

Section 2.5

2.37 We can use the ordered data in Figure 2.2 to determine the components of the box plot. To determine the quartiles, we note that $(0.25)(21) = 5.25$ and $(0.75)(21) = 15.75$.

For supplier A, $q_1 = 45 + (0.25)(45 - 45) = 45$ and $q_3 = 49 + (0.75)(49 - 49) = 49$. Thus, $IQR = 49 - 45 = 4$, $LOF = 45 - 3(4) = 33$, $LIF = 45 - (1.5)(4) = 39$, $UIF = 49 + (1.5)(4) = 55$, and $UOF = 49 + 3(4) = 61$. Since all values except 32 are between LIF and UIF, but 32 is less than LOF, the only outlier is 32. There are no suspect outliers. The MOSFET associated with the 32 hour time-to-failure should be considered carefully to gain insight into the unusually short lifetime.

For supplier B, $q_1 = 39 + (0.25)(40 - 39) = 39.25$ and $q_3 = 44 + (0.75)(45 - 44) = 44.75$. Thus, $IQR = 44.75 - 39.25 = 5.50$, $LOF = 39.25 - 3(5.50) = 22.75$, $LIF = 39.25 - (1.5)(5.50) = 31.00$, $UIF = 44.75 + (1.5)(5.50) = 53.00$, and $UOF = 44.75 + 3(5.50) = 61.25$. Since all values are between LIF and UIF, the data set for supplier B contains no suspect outliers and no outliers.

The medians for suppliers A and B are 46.5 and 42, respectively. Adding this information to that above, we obtain the grouped box plots (with lower fences for A indicated by dotted segments) given below. The MOSFETS obtained from supplier A tend to have the greater times to failure. However, those obtained from supplier B exhibit less overall variability in the times to failure. The empirical distribution for supplier B appears to be reasonably symmetric about the median.

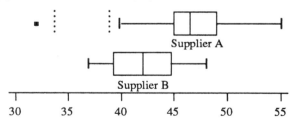

2.39 (a) For a data set of 12 measurements, we use $(0.25)(13) = 3.25$,
& $(0.50)(13) = 6.50$, and $(0.75)(13) = 9.75$ to find q_1, q_2, and q_3,
(b) respectively.

The ordered values for RF (the right front channel) are $x_{(1)} = x_{min} = 83.2$, $x_{(2)} = 83.6$, $x_{(3)} = 84.4$, $x_{(4)} = 85.6$, $x_{(5)} = 86.0$, $x_{(6)} = 86.8$, $x_{(7)} = 87.6$, $x_{(8)} = 90.4$, $x_{(9)} = 93.6$, $x_{(10)} = 94.0$, $x_{(11)} = 94.4$, and $x_{(12)} = 96.0$. Using these values, $q_1 = x_{(3)} + (0.25)(x_{(4)} - x_{(3)}) = 84.4 + (0.25)(85.6 - 84.4)$

$= 84.7, q_2 = x_{(6)} + (0.50)(x_{(7)} - x_{(6)}) = 86.8 + (0.50)(87.6 - 86.8) = 87.2$, and $q_3 = x_{(9)} + (0.75)(x_{(10)} - x_{(9)}) = 93.6 + (0.75)(94.0 - 93.6) = 93.9$. Since $IQR = 93.9 - 84.7 = 9.2$, the fences are $LOF = 84.7 - (3.0)(9.2) = 57.1$, $LIF = 84.7 - (1.5)(9.2) = 70.9$, $UIF = 93.9 + (1.5)(9.2) = 107.7$, and $UOF = 93.9 + (3.0)(9.2) = 121.5$. All values fall between the two inner fences, so there are no outliers or suspect outliers.

The quartiles and fences (to the nearest tenth) for the four channels are summarized in the following table. Those for the left front channel are from the solution of problem 2.36.

Channel	x_{min}	LOF	LIF	q_1	q_2	q_3	UIF	UOF	x_{max}
RF	83.2	57.1	70.9	84.7	87.2	93.9	107.7	121.5	96.0
RR	84.0	67.8	76.7	85.7	87.2	91.6	100.5	109.5	95.6
LF	82.8	63.2	74.0	84.8	87.0	92.0	102.8	113.6	108.8
LR	80.8	56.3	69.8	83.3	87.8	92.3	105.8	119.3	95.2

MYSTAT grouped box plots for the four channels are depicted in the following figure. Other than the suspect outlier associated with the left front channel (108.8), the distributions appear to be quite similar. Larger samples might help in detecting differences that are missed with such small samples.

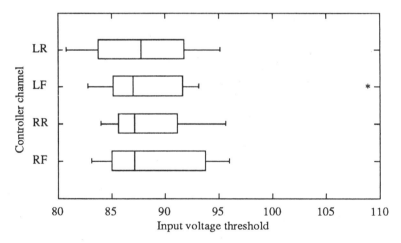

(c)   See problem 1.19 for a multi-vari chart. From that chart, we notice that the variability within controller 1 is much greater than that of the other controllers. Otherwise, the variabilities within controllers are similar. This information is not available with the grouped box plots.

The large LF value for controller 1 is the outlier identified on the box plot. Thus, a careful study of the LF channel of controller 1 should be conducted.

11

2.41 **(a)** A stem-and-leaf display is given to the right. No unusually large or small values are evident. I would search for a reason for the large numbers of absences in the upper thirty- and forty-day brackets. Perhaps some type of personal-day policy, or something associated with the particular days involved, could be contributing to those absences. A plot of the data in time-series order may provide some insight.

1*	2
1•	67
2*	012234
2•	666799
3*	12223
3•	5555677889
4*	01123
4•	5567778889
5*	23
5•	557

(b) Since $(0.25)(51) = 12.75$, $(0.50)(51) = 25.50$, and $(0.75)(51) = 38.25$,

$q_1 = x_{(12)} + (0.75)(x_{(13)} - x_{(12)}) = 26 + (0.75)(27 - 26) = 26.75$,
$q_2 = x_{(25)} + (0.50)(x_{(26)} - x_{(25)}) = 36 + (0.50)(37 - 36) = 36.50$, and
$q_3 = x_{(38)} + (0.25)(x_{(39)} - x_{(38)}) = 46 + (0.25)(47 - 46) = 46.25$. Thus,

$IQR = 46.25 - 26.75 = 19.50$, $LIF = 26.75 - (1.50)(19.50) = -2.50$, and
$UIF = 46.25 + (1.50)(19.50) = 75.50$. A box plot follows.

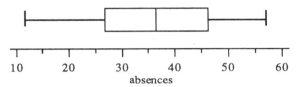

Since all data fall between the inner fences, there are no outliers or suspect outliers. The ranges of the four subsets determined by the quartiles are approximately equal.

2.43 Grouped box plots follow. With the 2 pyrometer system, the middle 50% of the data is less variable. The suspect outliers should be investigated to see if a measurement or recording error occurred. If so, the data should be corrected and an analysis of the revised data should follow. If not, a search for a cause should be conducted. If further investigation reveals that the values are probably correct, we have reason to believe the median preheat temperature has increased.

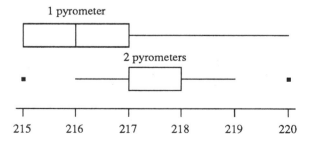

2.45 Using the digidot plot that follows, the empirical distribution is reasonably uniform about the median. The sample is too small to infer that the same is true about the population of fill weights.

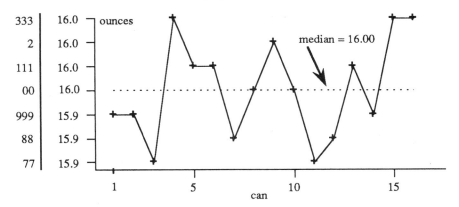

2.47 These data have a mild skew to the left. Since so many temperatures of 218 degrees were obtained, the number of temperatures below the median line is more than twice the number above. This is probably due to a measurement system that only records temperatures to the nearest degree. The time-series plot has a "saw tooth" appearance that may signal the presence of a special cause of much of the variability. For example, were all boards the same size and type?

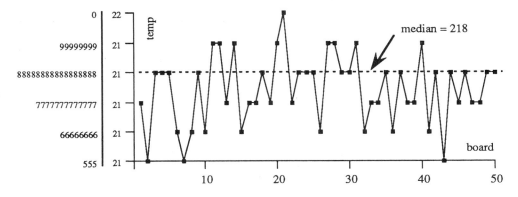

2.49 A scatter plot of the data pairs in problem 1.18, with the line $y = x$ through the origin included, follows. The S-shape formed by the points may signal data obtained from two or more populations. That is the case, since the points in the lower left-hand corner of the graphic were obtained after an increase in the hold pressure. It appears that the increase in pressure reduced the amount of shrink in both molds.

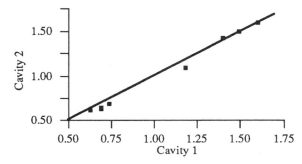

2.51 The extreme point in the lower-left portion of the Q-Q plot below is associated with the suspect outlier (32) for supplier A. If that point is ignored, the remainder of the graph seems to take a mild, "exponential" shape, indicating that times to failure for A vary more at the upper tail than do those for B. All points lie above the line $y = x$ (the dotted line on the graph), indicating that the median time to failure for supplier A is greater than that for B.

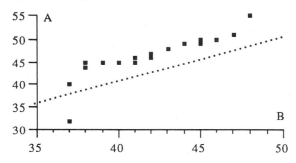

Section 2.9

2.53 **(a)** A scatter diagram is given in (d).

(b)

i	x_i	y_i	$x_i - \bar{x}$	$y_i - \bar{y}$	$(x_i - \bar{x})^2$	$(y_i - \bar{y})^2$	$(x_i - \bar{x})(y_i - \bar{y})$
1	0	3	-7/2	-16/3	49/4	256/9	112/6
2	1	5	-5/2	-10/3	25/4	100/9	50/6
3	3	8	-1/2	-1/3	1/4	1/9	1/6
4	4	7	1/2	-4/3	1/4	16/9	-4/6
5	6	13	5/2	14/3	25/4	196/9	70/6
6	7	14	7/2	17/3	49/4	289/9	119/6
total	21	50			150/4	858/9	348/6
					(SSX)	(SSY)	(SPXY)

(c) To avoid round-off error, the quantities in the summary table were calculated using fractions. Using the column totals, $\bar{x} = \sum x_i/6 = 21/6 = 7/2$, $\bar{y} = \sum y_i/6 = 50/6 = 25/3$, $SSX = \sum (x_i - \bar{x})^2 = 150/4 = 75/2 = 37.50$, $SSY = 858/9 = 286/3 \approx 95.33$, and $SPXY = \sum (x_i - \bar{x})(y_i - \bar{y}) = 348/6 = 58.00$. From Equations (2.12) and (2.13),

$b_1 = SPXY/SSX = 58 \div (75/2) = 116/75 \approx 1.55$ and $b_0 = \bar{y} - b_1 \bar{x} =$ $(25/3) - (116/75)(7/2) = 219/75 = 73/25 = 2.92$. The least squares line has equation $\hat{y} = 2.92 + 1.55x$.

(d) A scatter plot with the least squares line added follows. The line fits the data well, and there appears to be a strong linear relationship between the two variables.

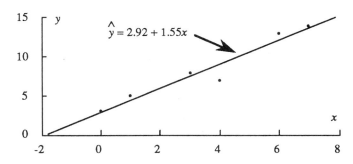

2.55 **(a), (b), & (c)** Using MYSTAT, the following scatter plot, least squares line, and least squares equation were obtained. The line is a good fit to the data. If a time of (say) 0.90 units is quoted to a customer, management can predict that $-0.027 + (1.084)(0.90) \approx 0.95$ units will be required.

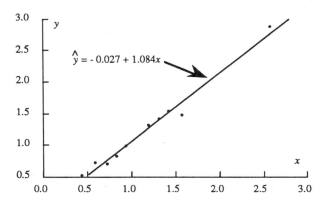

(d) The estimated values, residuals, and squares of the residuals are included in the following table. Those values were obtained using MYSTAT. $SSR \approx 0.030 + 0.057 = 0.087$ is the minimum value of the sum of the squared vertical deviations of the plotted points from a line. The sum of the squared vertical deviations of those points from any other line will be greater than 0.087.

15

i	x_i	y_i	$\hat{y}_i$	$y_i - \hat{y}_i$	$\left(y_i - \hat{y}_i\right)^2$	i	x_i	y_i	$\hat{y}_i$	$y_i - \hat{y}_i$	$\left(y_i - \hat{y}_i\right)^2$
1	0.58	0.72	0.602	0.118	0.014	6	2.58	2.87	2.769	0.101	0.010
2	0.43	0.51	0.439	0.071	0.005	7	1.19	1.30	1.263	0.037	0.001
3	1.42	1.54	1.512	0.028	0.001	8	0.94	0.98	0.992	−0.012	0.000
4	0.83	0.81	0.873	−0.063	0.004	9	1.57	1.46	1.674	−0.214	0.046
5	0.73	0.69	0.764	−0.074	0.006	10	1.31	1.40	1.393	0.007	0.000
					0.030						0.057

Section 2.10

2.57 **(a)** Using the results of Problem 2.53, $SSY = \Sigma(y_i - \bar{y})^2 = 858/9 = 286/3$, $SSX = 75/2$, and $SPXY = 58$. Thus, $r = 58/\sqrt{(75/2)(286/3)} \approx 0.97$ by Equation (2.17).

(b) From (a), $r^2 = (58)^2/(75/2)(286/3) = (58)^2/(25)(143) \approx 0.94$. The least squares line accounts for about 94% of the variability in the y-values.

2.59 Using prb02_55, $r \approx 0.990$ and $r^2 \approx 0.979$. The least squares line accounts for about 98% of the variability in the completed job times.

Supplementary Problems

2.61 **(a)** $\sum_{i=1}^{5} x_i = 2 + 6 + 4 + 0 + 3 = 15$

(b) $\sum_{i=1}^{5} x_i^2 = 2^2 + 6^2 + 4^2 + 0^2 + 3^2 = 4 + 36 + 16 + 0 + 9 = 65$

(c) $\sum_{i=1}^{5} (x_i - 3) = (2 - 3) + (6 - 3) + (4 - 3) + (0 - 3) + (3 - 3) = 0$

(d) $\sum_{i=1}^{5} (x_i - 3)^2 = (-1)^2 + (3)^2 + (1)^2 + (-3)^2 + (0)^2 = 1 + 9 + 1 + 9 + 0 = 20$

(e) Using (a), $\left(\sum_{i=1}^{5} x_i\right)^2 / 5 = 15^2/5 = 45$.

(f) Using (b) and (e), $\sum_{i=1}^{5} x_i^2 - \left[\left(\sum_{i=1}^{5} x_i\right)^2 / 5\right] = 65 - 45 = 20$.

2.63 $\sum_{i=1}^{n} (x_i - \bar{x}) = \sum_{i=1}^{n} x_i - \sum_{i=1}^{n} \bar{x} = n\bar{x} - n\bar{x} = 0$

2.65 $\sum_{i=1}^{30} x_i = 30 + 28 + 26 + \ldots + 29 + 28 + 30 = 877$

$\sum_{i=1}^{30} x_i^2 = 30^2 + 28^2 + 26^2 + \ldots + 29^2 + 28^2 + 30^2 = 25{,}713$

$SSX = 25{,}713 - \dfrac{877^2}{30} \approx 25{,}713 - 25{,}637.63 = 75.37$

$$s^2 = SSX/29 \approx 75.37/29 \approx 2.6$$

2.67 $\bar{x} = \dfrac{\sum_{i=1}^{4} x_i}{4} = \dfrac{250}{4} = 62.5$

$SSX = \sum_{i=1}^{4} x_i^2 - \dfrac{\left(\sum_{i=1}^{4} x_i\right)^2}{4} = 15{,}710 - \dfrac{250^2}{4} = 15{,}710 - 15{,}625 = 85$

$s^2 = SSX/3 = 85/3 \approx 28.33$ and $s = \sqrt{85/3} \approx 5.3$

2.69 **(a)** Using a computer and the 86 values remaining after deleting the outlier 97 from the data in prb02_01, $\bar{x} \approx 47.8$, $s^2 \approx 4.309$, and $s \approx 2.1$.

(b) The standardized value of 50 is approximately $z = (50 - 47.8)/2.1 \approx 1.05$. Thus, 50 is about 1.05 standard deviations above the mean.

2.71 **(a)** We first delete the obvious outlier, 491. For the remaining data, the range is $R = x_{max} - x_{min} = 406 - 375 = 31$. Using an interval width of 5 (which is near $R/7 = 31/7 \approx 4.43$) and beginning with 374.5 (which is below $x_{min} = 375$), gives the 7 class intervals in the following table of relative and relative cumulative frequencies. The relative frequency histogram is displayed below the table.

class	tally	f	rf	cf	rcf
[374.5, 379.5]	IIIII	5	0.085	5	0.085
[379.5, 384.5]	IIIII III	8	0.136	13	0.220
[384.5, 389.5]	IIIII I	6	0.102	19	0.322
[389.5, 394.5]	IIIII IIIII IIIII I	16	0.271	35	0.593
[394.5, 399.5]	IIIII IIIII IIII	14	0.237	49	0.831
[399.5, 404.5]	IIIII II	7	0.119	56	0.949
[404.5, 409.5]	III	3	0.051	59	1.000

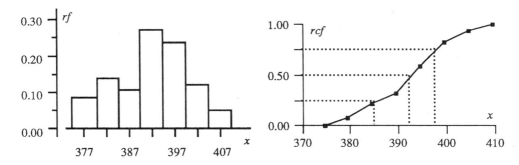

(b) Using the relative cumulative frequency distribution in the preceding table gives the ogive in the right-most figure above.

17

(c) Using the preceding ogive, the lower, middle, and upper quartiles are approximately 385, 392, and 398, respectively.

2.73 (a) Since 59 values remain after deleting the outlier, use $(0.25)(59+1) = 15$, $(0.50)(60) = 30$, and $(0.75)(60) = 45$ to determine the ordered values required when finding the quantiles. Using the stem-and-leaf display in Problem 2.3, we find that $q_1 = x_{(15)} = 387$, $\tilde{x} = x_{(30)} = 392$, and $q_3 = x_{(45)} = 397$. Thus, $IQR = q_3 - q_1 = 397 - 387 = 10$.

(b) Using $q_1 = 387$, $q_3 = 397$, and $IQR = 10$, $LOF = 387 - 3(10) = 357$, $LIF = 387 - (1.5)(10) = 372$, $UIF = 397 + (1.5)(10) = 412$, and $UOF = 397 + 3(10) = 427$. The only value beyond an outer fence is 491. That value is so much greater than UOF that our decision to delete it from further analyses seems warranted. All remaining values are between the inner fences, so there are no suspect outliers.

A box plot of the data (after excluding 491) is given below. The middle 50% of the empirical distribution appears to be reasonably symmetric.

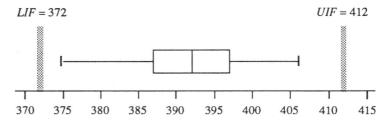

$LIF = 372$ $UIF = 412$

370 375 380 385 390 395 400 405 410 415

2.75 Using a computer and prb02_06, the lower, middle, and upper quartiles are 55, 57, and 60, respectively. Thus, $IQR = 5$, $LOF = 55 - 15 = 40$, $LIF = 55 - 7.5 = 47.5$, $UIF = 60 + 7.5 = 67.5$, and $UOF = 60 + 15 = 75$. Since $x_{min} = 49$ and $x_{max} = 68$, there are no outliers. The only suspect outlier is $x_{max} = 68$. If the frequency distribution and relative frequency histogram of Problem 2.6 are considered, that suspect outlier does not appear to be due to anything other than chance.

A box plot is presented below. The empirical distribution appears to be slightly skewed to the right.

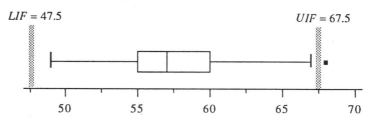

$LIF = 47.5$ $UIF = 67.5$

50   55   60   65   70

2.77   Using a computer and prb02_07, the lower, middle, and upper quartiles are 4, 4.5, and 4.5, respectively. Thus, $IQR = 0.5$, $LOF = 4 - 1.5 = 2.5$, $LIF = 4 - 0.75 = 3.25$, $UIF = 4.5 + 0.75 = 5.25$, and $UOF = 4.5 + 1.5 = 6.0$. One value, 3.0, lies between the lower fences, and 6 values (all 5.5) lie between the upper fences. Thus, there are 7 suspect outliers and no outliers. A box plot, with this information added, follows. The lower 50% of the distribution appears to be more variable than the upper 50%.

Note that only 6 distinct lever efforts were observed: 3.0, 3.5, 4.0, 4.5, 5.0, and 5.5. This is, most likely, due to an inadequate measurement system.

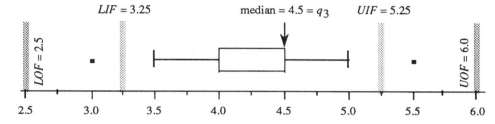

2.79   **(a)**   Using a computer and prb02_09, $\bar{x} \approx 3.36$, $s^2 \approx 2.6546$, and $s \approx 1.63$.
   (b)   The standardized value of 8.4 is approximately $(8.40 - 3.36)/1.63 \approx 3.09$. Thus, 8.4 is approximately 3.09 standard deviations above the mean.

2.81 $\sum_{i=1}^{15} x_i f_i = (-3)(1) + (-1)(2) + \ldots + (11)(4) + (12)(5) = 4223$

   $\sum_{i=1}^{15} x_i^2 f_i = (-3)^2(1) + (-1)^2(2) + \ldots + (11)^2(4) + (12)^2(5) = 28,087$

   $SSX = 28,087 - 4223^2/718 \approx 3248.9373$

   $s^2 = SSX/717 \approx 3248.9373/717 \approx 4.5313$

   $s = \sqrt{s^2} \approx \sqrt{4.5313} \approx 2.13$

2.83   **(a)** $\sum_{i=1}^{5} M_i f_i = (7)(8) + (10)(27) + (13)(35) + (16)(26) + (19)(4) = 1273$

   $\bar{M} = 1273/100 = 12.73 \approx 12.7$

   $\sum_{i=1}^{5} M_i^2 f_i = (7^2)(8) + (10^2)(27) + (13^2)(35)$
   $+ (16^2)(26) + (19^2)(4) = 17,107$

   $SSM = 17,107 - 1273^2/100 = 901.71$

   $s_M^2 = 901.71/99 \approx 9.11$ and $s_M = \sqrt{901.71/99} \approx 3.0$

   (b)   When rounded to the nearest tenth, the corresponding means and standard deviations are equal.

(c)

Interval	M	f	rf	cf	rcf
5.5 - 8.5	7	8	0.08	8	0.08
8.5 - 11.5	10	27	0.27	35	0.35
11.5 - 14.5	13	35	0.35	70	0.70
14.5 - 17.5	16	26	0.26	96	0.96
17.5 - 20.5	19	4	0.04	100	1.00

Using the preceding summary, the following ogive is obtained. From the ogive, the lower, middle, and upper quartiles are approximately 10, 13, and 15 pound-inches, respectively. Using the 100 individual values, $q_1 = 11$, $q_2 = 12$, and $q_3 = 15$.

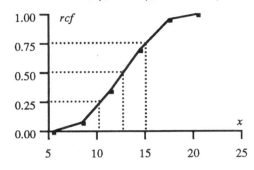

2.85 **(a)** Letting $y = (M - 0.830)(250)$ denote the coded value of the midpoint M gives the following coded values and frequencies.

y	-4	-3	-2	-1	0	1	2	3	4
f	1	2	3	5	6	10	8	4	1

(b) $\bar{y} = \dfrac{\sum_{i=1}^{9} y_i f_i}{40} = [(-4)(1) + (-3)(2) + \ldots + (3)(4) + (4)(1)]/40$

$$= 21/40 = 0.525,$$

$$SSY = \sum_{i=1}^{9} y_i^2 f_i - \frac{\left(\sum_{i=1}^{9} y_i f_i\right)^2}{40}$$

$$= [(-4)^2(1) + (-3)^2(2) + \ldots + (3)^2(4) + (4)^2(1)] - \frac{21^2}{40}$$

$$= 145 - 11.025 = 133.975$$

$$s_Y^2 = SSY/39 = 133.975/39 \approx 3.43526$$

$$s_Y = \sqrt{133.975/39} \approx 1.853.$$

(c) If c_1 and c_2 are constants, Equations (2.1) and (2.5) can be used to show that $\overline{M} = c_1 \bar{y} + c_2$ and $SSM = \left(c_1^2\right)(SSY)$ when $M = c_1 y + c_2$. Since $M = \frac{1}{250} y + 0.830$, we let $c_1 = 1/250$, $c_2 = 0.830$, $\bar{y} = 0.525$, and

$SSY = 133.975$ to obtain:

- $\overline{M} = \frac{1}{250}\overline{y} + 0.830 = \frac{1}{250}(0.525) + 0.830 = 0.8321$
- $SSM = \left(\frac{1}{250}\right)^2 (SSY) = \left(\frac{1}{250}\right)^2 (133.975) = 0.0021436$
- $s_M = \sqrt{(SSM)/39} = \sqrt{0.0021436/39} \approx 0.0074$.

2.87
$$
\begin{aligned}
SPXY &= \sum_{i=1}^{n}(x_i - \overline{x})(y_i - \overline{y}) \\
&= \sum_{i=1}^{n}[x_i y_i - (\overline{x})y_i - (\overline{y})x_i + (\overline{x})(\overline{y})] \\
&= \sum_{i=1}^{n}x_i y_i - \overline{x}\sum_{i=1}^{n}y_i - \overline{y}\sum_{i=1}^{n}x_i + n(\overline{x})(\overline{y}) \\
&= \sum_{i=1}^{n}x_i y_i - \overline{x}(n\overline{y}) - \overline{y}(n\overline{x}) + n(\overline{x})(\overline{y}) \\
&= \sum_{i=1}^{n}x_i y_i - n(\overline{x})(\overline{y}) \\
&= \sum_{i=1}^{n}x_i y_i - n\left(\frac{\sum_{i=1}^{n}x_i}{n}\right)\left(\frac{\sum_{i=1}^{n}y_i}{n}\right) \\
&= \sum_{i=1}^{n}x_i y_i - \frac{\left(\sum_{i=1}^{n}x_i\right)\left(\sum_{i=1}^{n}y_i\right)}{n}
\end{aligned}
$$

2.89
$$
\begin{aligned}
(SSY)(1 - r^2) &= (SSY)\left(1 - \frac{(SPXY)^2}{(SSX)(SSY)}\right) \\
&= SSY - \frac{(SPXY)^2}{SSX} \\
&= \sum_{i=1}^{n}(y_i - \overline{y})^2 - \sum_{i=1}^{n}(\hat{y}_i - \overline{y})^2 \text{, by Equation (2.21)} \\
&= SSR \text{, by Equation (2.20)}
\end{aligned}
$$

Using the data in Problem 2.53 and a calculator, $r^2 \approx 0.940979$. From Problem 2.88, $SSY = 286/3$. Thus, $SSR \approx (286/3)(1 - 0.940979) \approx 5.63$.

CHAPTER 3

PROBABILITY AND PROBABILITY FUNCTIONS

Section 3.1

3.1 (a) $S = \{SSS, SSF, SFS, SFF, FSS, FSF, FFS, FFF\}$ (b) $E = \{SSS\}$

3.3 Assuming outcomes are equally likely, $P(E) = 8/(10 + 8) = 4/9 \approx 0.44$.

3.5 Since $2 + 7 = 9$ of the $2 + 5 + 4 + 7 = 18$ bulbs are red or green,
$P(\text{red or green}) = 9/18 = 0.5$.

3.7 Since $1 = p_1 + p_2 + p_3 + p_4 = 0.2 + 0.1 + 0.5 + p_4 = 0.8 + p_4$, $p_4 = 0.2$.

3.9 Since $p_3 = 1 - (p_1 + p_2 + p_4 + p_5 + p_6) = 1 - 0.94 = 0.06$, the probability that
the housing was produced on a machine made by an Ohio company is
$p_1 + p_3 = 0.26 + 0.06 = 0.32$.

Section 3.2

3.11 (a) $P(\text{male} \mid 30 \text{ or older}) = 200/650 = 4/13 \approx 0.31$
     (b) $P(\text{female and under } 30) = 100/1{,}000 = 0.10$
     (c) $P(30 \text{ or older} \mid \text{female}) = 450/550 = 9/11 \approx 0.82$
     (d) $P(\text{female or at least } 30) = (550 + 200)/1{,}000 = 0.75$
     (e) $P(\text{under } 30) = 350/1{,}000 = 0.35$
     (f) $P(\text{under } 30 \mid \text{male}) = 250/450 = 5/9 \approx 0.56$

3.13 (a) $5/49 \approx 0.1020$
     (b) $3/48 = 1/16 = 0.0625$
     (c) $4/48 = 1/12 \approx 0.0833$

Section 3.3

3.15 Let M_i denote "a defective item is missed by the ith inspector." The
probability that a defective item is missed by both inspectors is
$P(M_1 \cap M_2) = P(M_1)P(M_2 \mid M_1) = (0.10)(0.50) = 0.05$.

3.17 $(5/50)(4/49) = 2/245 \approx 0.0082$

3.19 Since each switch works properly with probability 0.98, the probability
that all four function properly is $(0.98)^4 \approx 0.9224$.

3.21 $(46/50)(45/49)(44/48)(43/47)(42/46) = 4{,}257/6{,}580 \approx 0.6470$

3.23 $(0.90)(0.85)(0.98) = 0.7497$

3.25

outcome	probability	outcome	probability
SSS	$(0.90)(0.85)(0.98) = 0.7497$	FSS	$(0.10)(0.85)(0.98) = 0.0833$
SSF	$(0.90)(0.85)(0.02) = 0.0153$	FSF	$(0.10)(0.85)(0.02) = 0.0017$
SFS	$(0.90)(0.15)(0.98) = 0.1323$	FFS	$(0.10)(0.15)(0.98) = 0.0147$
SFF	$(0.90)(0.15)(0.02) = 0.0027$	FFF	$(0.10)(0.15)(0.02) = 0.0003$

Since the eight probabilities sum to 1 and the parallel system functions properly in all cases except FFF, the probability that the parallel system functions properly is $1 - P(\text{FFF}) = 1 - 0.0003 = 0.9997$.

3.27 $(0.98)(0.95)(0.99) = 0.92169$

3.29 $P(A \cup B \cup C) = P(A) + P(B) + P(C) - P(A \cap B) - P(A \cap C) - P(B \cap C)$
$+ P(A \cap B \cap C)$

$= 0.20 + 0.16 + 0.14 - 0.08 - 0.05 - 0.04 + 0.02 = 0.35$

3.31

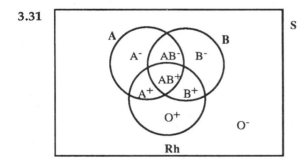

3.33 **(a)** Let $M_1 \cap M_2$ denote "an illegal lamp is missed by the first inspection and the second inspection." From the multiplicative law, $P(M_1 \cap M_2) = P(M_1)P(M_2 \mid M_1) = (0.36)(0.15) = 0.054$.

(b) Let C_i denote "an illegal lamp is found on the ith inspection." The probability that the inspection system catches an illegal lamp is $P(C_1) + P(M_1 \cap C_2) = P(C_1) + P(M_1)P(C_2 \mid M_1) = 0.64 + (0.36)(0.85) = 0.946$.

3.35 Let C denote "the selected coil conforms to specifications" and N denote "the selected coil is nonconforming." For a very large run, the outcomes at each selection can be treated as independent. Under this assumption, we obtain the probabilities on the following stochastic tree. Using that tree, the probability of accepting a run for which 3% of the coils are nonconforming is $(0.97)^2 + 2(0.03)(0.97)^2 + (0.03)^2(0.97) = 0.998227 \approx 0.998$.

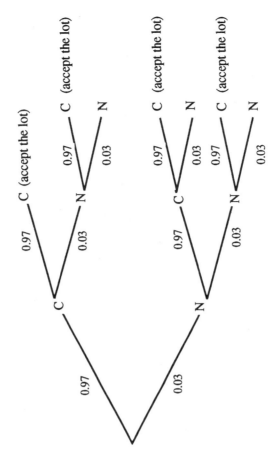

3.37 Let M_i denote "the part was produced on the ith machine"; $i = 1, 2, 3, 4, 5$. Let S_0 and S_1 denote "the part does not have excessive shrink" and "the part has excessive shrink," respectively. Using these notations and the probabilities given in the problem statement, we obtain the following stochastic tree.

(a) $P(M_4) = 0.20$ **(b)** $P(S_1 \mid M_3) = 0.08$

(c) Using the following stochastic tree, the probability that a randomly chosen lens will have excessive shrink is

$$
\begin{aligned}
P(S_1) &= P(M_1 \cap S_1) + P(M_2 \cap S_1) + P(M_3 \cap S_1) + P(M_4 \cap S_1) \\
&\quad + P(M_5 \cap S_1) \\
&= (0.20)(0.05) + (0.15)(0.06) + (0.10)(0.08) + (0.20)(0.05) \\
&\quad + (0.35)(0.03) = 0.0475
\end{aligned}
$$

(d) Using (c), $P(M_2 \mid S_1) = P(M_2 \cap S_1) / P(S_1)$
$$= (0.15)(0.06)/0.0475 = 18/95 \approx 0.1895.$$

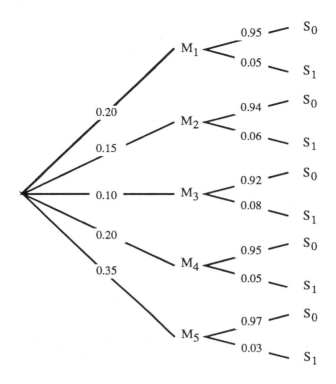

3.39 Let L denote "the chief engineer is late," E denote "the ticket is purchased from airline A," F denote "the ticket is purchased from airline B," and G denote "the ticket is purchased from airline C." Since $E \cap L$, $F \cap L$, and $G \cap L$ are pairwise mutually exclusive,

$$
\begin{aligned}
P(L) &= P(E \cap L) + P(F \cap L) + P(G \cap L) \\
&= P(E)P(L \mid E) + P(F)P(L \mid F) + P(G)P(L \mid G) \\
&= (0.60)(0.01) + (0.30)(0.03) + (0.10)(0.05) \\
&= 0.02.
\end{aligned}
$$

Section 3.5

3.41 $P(\text{at least one is nonconforming}) = 1 - P(\text{all are conforming})$
$$= 1 - (0.88)^{20} \approx 0.92$$

3.43 $S = \{A_0B_0C_0, A_0B_0C_1, A_0B_1C_0, A_0B_1C_1, A_1B_0C_0, A_1B_0C_1, A_1B_1C_0, A_1B_1C_1\}$
$E_2 = \{A_1B_0C_1\}$
$E_2 = \{A_0B_0C_0, A_0B_0C_1, A_0B_1C_0, A_0B_1C_1, A_1B_0C_0, A_1B_1C_0, A_1B_1C_1\}$

3.45 If $E_2 =$ "the device operates and component B fails," then $E_2 = \{A_1B_0C_1\}$. Using the following stochastic tree, $P(E_2) = p_6 = 0.111672 \approx 0.11$.

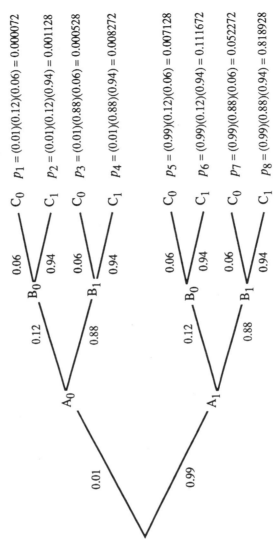

$p_1 = (0.01)(0.12)(0.06) = 0.000072$

$p_2 = (0.01)(0.12)(0.94) = 0.001128$

$p_3 = (0.01)(0.88)(0.06) = 0.000528$

$p_4 = (0.01)(0.88)(0.94) = 0.008272$

$p_5 = (0.99)(0.12)(0.06) = 0.007128$

$p_6 = (0.99)(0.12)(0.94) = 0.111672$

$p_7 = (0.99)(0.88)(0.06) = 0.052272$

$p_8 = (0.99)(0.88)(0.94) = 0.818928$

3.47 Let F_i denote "the ith component fails"; $i = 1, 2, 3$. Since $P(F_1 F_2 F_3) = (0.02)(0.05)(0.01) = 0.00001$, the probability that the parallel system functions properly is $1 - 0.00001 = 0.99999$.

3.49 Let E = "the coil has proper inductance" and F = "the coil has proper resistance." The probability that the selected coil has neither proper inductance nor proper resistance is $1 - P(E \cup F) = 1 - [P(E) + P(F) - P(E \cap F)] = 1 - [0.40 + 0.55 - (0.40)(0.50)] = 1 - 0.75 = 0.25$.

3.51 An intruder is detected if at least one of n devices is activated. This happens with probability $1 - (0.30)^n$. Since $1 - (0.30)^n \geq 0.97$ when $(0.30)^n \leq 0.03$, $(0.30)^2 = 0.09$, and $(0.30)^3 = 0.027$, 3 devices are required.

3.53 **(a)** The system is two simple series subsystems connected in parallel. Since each series subsystem fails with probability $1 - (0.75)(0.75) = 0.4375$, the system functions properly with probability $1 - (0.4375)^2 \approx 1 - 0.1914 = 0.8086$.

   (b) Let E = "the first series subsystem functions properly" and F = "the second series subsystem functions properly." From (a), $P(E) = P(F) = 1 - 0.4375 = 0.5625$ and $P(E') = P(F') = 0.4375$. But E and F are independent, so E and F' as well as E' and F are also independent. Thus, the probability that exactly one series subsystem fails is

$$P(E \cap F') + P(E' \cap F) = P(E)P(F') + P(E')P(F)$$
$$= 2(0.5625)(0.4375) \approx 0.4922.$$

   (c) A parallel system of n such series subsystems functions properly with probability $1 - (0.4375)^n$. If that probability is at least 0.99, then $(0.4375)^n \leq 0.01$. This implies that $n \geq [\ln(0.01)]/[\ln(0.4375)] \approx 5.57$. So, 6 such subsystems are required.

Section 3.6

3.55 $\dfrac{(5)(4)(24)(23)(22)(21)}{(26)(25)(24)(23)(22)(21)} = \dfrac{2}{65} \approx 0.0308$

3.57 **(a)** $C(9, 3)C(7, 2)/C(16, 5) = (84)(21)/4,368 = 21/52 \approx 0.4038$

   (b) $[C(9, 1)C(7, 4) + C(9, 0)C(7, 5)]/C(16, 5) = [(9)(35) + (1)(21)]/4,368$
$$= 1/13 \approx 0.0769$$

3.59 **(a)** $7^5 = 16,807$ **(b)** $P(7, 5) = 2,520$

3.61 $C(1, 1)C(49, 4)/C(50, 5) = 211,876/2,118,760 = 1/10 = 0.1$

3.63 $1 - C(47, 5)/C(50, 5) = 1 - (1,533,939/2,118,760)$
$$= 1 - (1,419/1,960)$$
$$= 541/1,960 \approx 0.2760$$

3.65 **(a)** The number of possible unordered samples is $C(12, 3) = 220$.
   [**Note:** There are $P(12, 3) = 1,320$ possible ordered samples.]

   (b) Let $o_1, o_2, \ldots, o_{11}, o_{12}$ denote the 12 items. To determine how many of each are included in the sample, think of arranging 11 vertical bars and 3 asterisks along a horizontal line segment that has a wall at each of its ends. The vertical bars and walls determine 12 "bins". Beginning with the leftmost bin, let the number of asterisks between the wall and the first vertical bar represent the number of occurrences of o_1, the number between the first and second vertical bars represent the number of occurrences of $o_2, \ldots$, and the number between the twelfth vertical bar and the rightmost wall represent the number of occurrences of o_{12}. For example,

$$| \ \ | \ \ | \ \ |*| \ \ | \ \ | \ \ | \ \ |**| \ \ | \ \ | \ \ |$$

indicates that the sample $\{o_4, o_9, o_9\}$ was obtained. Since there are

C(14, 3) = 364 different arrangements of the 11 vertical bars and 3 asterisks, there are 364 different samples of size 3 that can be obtained when sampling with replacement from a lot of 12 items.

Note: Generalizing the preceding discussion, when sampling with replacement from a lot of n items, there are $C(n + r - 1, r)$ possible unordered samples of size r.

Note: The number of ordered samples of size 3 possible when sampling with replacement from a lot of 12 items is $(12)(12)(12) = 1{,}728$. In general, there are n^r possible ordered samples of size r, when sampling with replacement from a lot of n items.

Section 3.7.1

3.67 **(a)**
$$F(x) = P(X \le x) = \begin{cases} 0.0, & x < 0 \\ 0.3, & 0 \le x < 1 \\ 0.9, & 1 \le x < 2 \\ 1.0, & 2 \le x \end{cases}$$

(b)

3.69 **(a)**
$$F(x) = P(X \le x) = \begin{cases} 0.0000, & x < 0 \\ 0.3830, & 0 \le x < 1 \\ 0.8336, & 1 \le x < 2 \\ 0.9838, & 2 \le x < 3 \\ 0.9996, & 3 \le x < 4 \\ 1.0000, & 4 \le x \end{cases}$$

(b)

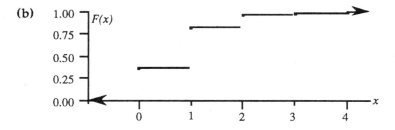

28

3.71 $P(X > 8) = f(10) + f(15) = 0.3 + 0.3 = 0.6$

3.73 For the function given, $\sum_{x=0}^{1} f(x) = f(0) + f(1) = 1 + 1 = 2$. Since this sum exceeds 1, f is not a discrete probability function.

3.75 $\begin{aligned} E[X] &= 1f(1) + 2f(2) + 3f(3) + 4f(4) + 5f(5) + 6f(6) \\ &= 1(0.1) + 2(0.1) + 3(0.1) + 4(0.1) + 5(0.1) + 6(0.5) \\ &= 4.5 \end{aligned}$

$\begin{aligned} Var(X) &= (1 - 4.5)^2 (0.1) + (2 - 4.5)^2 (0.1) + (3 - 4.5)^2 (0.1) + (4 - 4.5)^2 (0.1) \\ &\quad + (5 - 4.5)^2 (0.1) + (6 - 4.5)^2 (0.5) = 3.25 \end{aligned}$

3.77 $E[X] = (-3)(0.01) + (0)(0.25) + (1)(0.40) + (2)(0.30) + (7)(0.04) = 1.25$

$E[X^2] = (-3)^2(0.01) + (0^2)(0.25) + (1^2)(0.40) + (2^2)(0.30) + (7^2)(0.04) = 3.65$

$Var(X) = E[X^2] - (E[X])^2 = 3.65 - 1.25^2 = 3.65 - 1.5625 = 2.0875$

$\sigma = \sqrt{Var(X)} = \sqrt{2.0875} \approx 1.44$

3.79 The probability function for Problem 3.72 is given in the table to the right.

x	0	1	2	3
$f(x)$	0.90	0.05	0.02	0.03

$E[X] = 0(0.90) + 1(0.05) + 2(0.02) + 3(0.03) = 0.18$

$\begin{aligned} Var(X) &= (0 - 0.18)^2(0.90) + (1 - 0.18)^2(0.05) + (2 - 0.18)^2(0.02) \\ &\quad + (3 - 0.18)^2(0.03) = 0.3676 \end{aligned}$

3.81 Let X denote the profit per component. Values of X and the corresponding probabilities are summarized in the given table.

x	20	5	-10
$f(x)$	0.90	0.08	0.02

Using those results, the expected profit per component is $E[X] = (20)(0.90) + (5)(0.08) + (-10)(0.02) = \$18.20$.

3.83 If X denotes the number of accidents, $E[X] = 0(0.90) + 1(0.04) + 2(0.03) + 3(0.02) + 4(0.01) = 0.20$. In the long run, there are 2 such accidents every 10 Saturdays.

Section 3.7.2

3.85 **(a)** Since $f(x) \geq 0$ for $0 \leq x \leq 3$ and $\int_0^3 (2/9)x\,dx = (x^2/9)\Big]_0^3 = 1$, $f(x)$ is a continuous probability density function.

(b) $P(X > 2) = \int_2^3 (2/9)x\,dx = (x^2/9)\Big]_2^3 = 1 - (4/9) = 5/9$

(c) $\begin{aligned} P(0.50 < X \leq 1.50) &= \int_{0.50}^{1.50} \left(\frac{2}{9}x\right)dx \\ &= \left(x^2/9\right)\Big]_{0.50}^{1.50} \\ &= (2.25/9) - (0.25/9) = 2/9 \end{aligned}$

3.87 **(a)** Since $f(x) \geq 0$ for $0 \leq x \leq 1$ and $\int_0^1 \frac{3}{2}\sqrt{x}\,dx = x^{3/2}\Big]_0^1 = 1$, $f(x)$ is a continuous probability density function.

(b) $P(X \leq 0.5) = \int_0^{0.5} \frac{3}{2}\sqrt{x}\,dx = x^{3/2}\Big]_0^{0.5} = (0.5)^{3/2} = \sqrt{2}/4 \approx 0.3536$

(c) $E[X] = \int_0^1 x\left(\frac{3}{2}\right)\sqrt{x}\,dx = \frac{3}{2}\int_0^1 x^{3/2}\,dx = \frac{3}{5}x^{5/2}\Big]_0^1 = 3/5$

$$
\begin{aligned}
Var(X) &= E[X^2] - (E[X])^2 \\
&= \int_0^1 x^2\left(\frac{3}{2}\right)\sqrt{x}\,dx - \left(\frac{3}{5}\right)^2 \\
&= \frac{3}{2}\int_0^1 x^{5/2}\,dx - \left(\frac{3}{5}\right)^2 \\
&= \left[\frac{3}{7}x^{7/2}\right]_0^1 - \frac{9}{25} = \frac{3}{7} - \frac{9}{25} = \frac{12}{175} \approx 0.0686
\end{aligned}
$$

3.89 **(a)** If $3 \leq x$, $F(x) = P(3 \leq X \leq x) = \int_3^x [5(3^5)/t^6]\,dt = -(3/t)^5\Big]_3^x = 1 - (3/x)^5$. Thus,

$$
F(x) = \begin{cases} 0 & , \ x < 3 \\ 1-(3/x)^5, & 3 \leq x \end{cases}.
$$

(b)
$$
\begin{aligned}
E[X] &= \int_3^\infty x[5(3^5)/x^6]\,dx \\
&= \lim_{b \to \infty}\left[-\frac{15}{4}\left(\frac{3}{x}\right)^4\right]_3^b \\
&= \lim_{b \to \infty}\left[-\frac{15}{4}\left(\frac{3}{b}\right)^4 + \frac{15}{4}\right] = 15/4
\end{aligned}
$$

$$
\begin{aligned}
Var(X) &= E[X^2] - (E[X])^2 \\
&= \int_3^\infty x^2\left[5(3^5)/x^6\right]dx - \left(\frac{15}{4}\right)^2 \\
&= \lim_{b \to \infty}\left[15\int_3^b\left(\frac{3}{x}\right)^4 dx\right] - \left(\frac{15}{4}\right)^2 \\
&= \lim_{b \to \infty}\left[\frac{-405}{x^3}\right]_3^b - \frac{225}{16} \\
&= \lim_{b \to \infty}\left[\frac{-405}{b^3} + 15\right] - \frac{225}{16} \\
&= 15 - (225/16) = 15/16 = 0.9375
\end{aligned}
$$

3.91 **(a)** Using integration by parts or a table of integrals,

$$
\begin{aligned}
E[X] &= \int_{0.5}^\infty (0.15)xe^{-(0.15)(x-0.5)}\,dx \\
&= \lim_{b \to \infty}\left[-(20/3)(0.15x+1)e^{-(0.15)(x-0.5)}\right]_{0.5}^b \\
&= \lim_{b \to \infty}\left[\frac{(-20)(0.15b+1)}{3e^{0.15(b-0.5)}} + \frac{20(1.075)}{3}\right] = 0 + \frac{21.5}{3} = \frac{43}{6}
\end{aligned}
$$

30

$$\sigma_X^2 = E[X^2] - (E[X])^2$$

$$= \int_{0.5}^{\infty} (0.15)x^2 e^{-(0.15)(x-0.5)}dx - \left(\tfrac{43}{6}\right)^2$$

$$= -(e^{0.075}/(0.15)^2)\int_{0.5}^{\infty}(-0.15x)^2 e^{-0.15x}(-0.15)dx - \left(\tfrac{43}{6}\right)^2$$

$$= \left(-\tfrac{400}{9}\right)e^{0.075} \cdot \lim_{b\to\infty}\left[-0.15x^2 e^{-0.15x} + 2\int 0.15xe^{-0.15x}(-0.15)dx\right]_{0.5}^{b} - \left(\tfrac{43}{6}\right)^2$$

$$= \left(-\tfrac{400}{9}\right)e^{0.075} \cdot \lim_{b\to\infty}\left[-0.15x^2 e^{-0.15x} + 0.30x + 2e^{-0.15x}\right]_{0.5}^{b} - \left(\tfrac{43}{6}\right)^2$$

$$= \lim_{b\to\infty}\left[-\left(\frac{9x^2+120x+800}{9}\right)e^{-(0.15)(x-0.5)}\right]_{0.5}^{b} - \left(\tfrac{43}{6}\right)^2$$

$$= \lim_{b\to\infty}\left[-\left(\frac{9b^2+120b+800}{9e^{(0.15)(b-0.5)}}\right) + \frac{862.25}{9}\right] - \left(\tfrac{43}{6}\right)^2$$

$$= [0 + (3{,}449/36)] - (1{,}849/36) = 400/9$$

(b) $F(2) = P(X \le 2) = \int_{0.5}^{2}(0.15)e^{-(0.15)(x-0.5)}dx$

$$= \left[-e^{-(0.15)(x-0.5)}\right]_{0.5}^{2}$$

$$= 1 - e^{-0.225} \approx 0.2015$$

(c) If $x \ge 0.5$, $F(x) = P(0.5 \le X \le x) = \int_{0.5}^{x}(0.15)e^{-(0.15)(t-0.5)}dt$

$$= \left[-e^{-(0.15)(t-0.5)}\right]_{0.5}^{x}$$

$$= 1 - e^{-0.15(x-0.5)}. \text{ Thus,}$$

$$F(x) = \begin{cases} 0 & , \ x < 0.5 \\ 1 - e^{-(0.15)(x-0.5)} & , \ 0.5 \le x \end{cases}$$

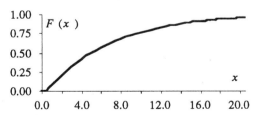

Section 3.74

3.93 **(a)** Parts are assembled randomly; so, we will assume statistical independence. Since $\mu_Y = \mu_1 + \mu_2 + \mu_3 = 1.500 + 1.000 + 0.500 = 3.000$ inches, and

$$\sigma_Y = \sqrt{\left(\tfrac{0.005}{3}\right)^2 + \left(\tfrac{0.004}{3}\right)^2 + \left(\tfrac{0.003}{3}\right)^2} = \tfrac{1}{3}\sqrt{0.000050} \text{ inch,}$$

$$\mu_Y \pm 3\sigma_Y = 3.000 \pm \sqrt{0.000050} \approx 3.000 \pm 0.007.$$

(b) Suppose $Y = X_1 + X_2 + X_3$ has natural tolerances of $\mu_Y \pm 3\sigma_Y$. Also, suppose $\sigma_2 = \frac{4}{5}\sigma_1$ and $\sigma_3 = \frac{3}{5}\sigma_1$. Since

$$\sigma_Y^2 = \sigma_1^2 + \sigma_2^2 + \sigma_3^2 = \sigma_1^2 + \tfrac{16}{25}\sigma_1^2 + \tfrac{9}{25}\sigma_1^2 = 2\sigma_1^2,$$

$\sigma_1 = (\sqrt{2}/2)\sigma_Y$, $\sigma_2 = (2\sqrt{2}/5)\sigma_Y$, and $\sigma_3 = (3\sqrt{2}/10)\sigma_Y$. Thus, $\mu_1 \pm \frac{3\sqrt{2}}{2}\sigma_Y$, $\mu_2 \pm \frac{6\sqrt{2}}{5}\sigma_Y$, and $\mu_3 \pm \frac{9\sqrt{2}}{10}\sigma_Y$ are the recommended tolerances.

Supplementary Problems

3.95 (a) $P(A \cap B) = P(A)P(B) = (1/2)(1/3) = 1/6$
    (b) $P(A \mid B) = P(A) = 1/2$
    (c) $P(A \cup B) = P(A) + P(B) - P(A \cap B) = (1/2) + (1/3) - (1/6) = 2/3$

3.97 Let p denote the probability that a component functions properly. Assuming independence, $p^2 = 0.90$. Thus, $p = \sqrt{0.90} \approx 0.95$.

3.99 (a) $(1 - 0.025)(1 - 0.011) = (0.975)(0.989) = 0.964275 \approx 0.96$
    (b) $1 - 0.964275 = 0.035725 \approx 0.04$

3.101 $(1 - 0.40)(1 - 0.20)(1 - 0.10) = (0.60)(0.80)(0.90) = 0.432$

3.103 The probability that all engines fail is $(0.05)^4$ and the probability that exactly 3 engines fail is $4(0.05)^3(0.95)$. Thus, the probability of a crash is $(0.05)^4 + 4(0.05)^3(0.95) = 0.00048125 \approx 0.0005$.

3.115 (a) $P(D) = 0.02$, $P(D') = 0.98$, $P(C \mid D) = 0.97$, $P(C \mid D') = 0.05$

    (b) $P(D \mid C) = \dfrac{P(D)P(C \mid D)}{P(D)P(C \mid D) + P(D')P(C \mid D')}$

           $= \dfrac{(0.02)(0.97)}{(0.02)(0.97) + (0.98)(0.05)}$

           $= 194/684 \approx 0.2836$

    (c) $P(C) = (0.02)(0.97) + (0.98)(0.05) = 0.0684$

3.117 Let A = "attended the seminar" and T = "uses the techniques."
$P(T) = P(A)P(T \mid A) + P(A')P(T \mid A') = (0.80)(0.89) + (0.20)(0.42) = 0.7960$

CHAPTER 4

BINOMIAL, POISSON, AND HYPERGEOMETRIC DISTRIBUTIONS

Section 4.1

4.1 The space of X is $S = \{9, 18\}$.
$P(X = 9) = f(9) = \theta$ and $P(X = 18) = f(18) = 1 - \theta$.
$\mu = (9)(\theta) + (18)(1 - \theta) = 9(2 - \theta)$
$$\begin{aligned}\sigma^2 &= [9 - 9(2 - \theta)]^2(\theta) + [18 - 9(2 - \theta)]^2(1 - \theta) \\ &= 81(\theta - 1)^2(\theta) + 81(\theta^2)(1 - \theta) = 81(\theta)[(\theta - 1)^2 + \theta(1 - \theta)] = 81(\theta)(1 - \theta)\end{aligned}$$

4.3 Letting X denote the number of defective parts in the sample of 10 parts, we assume that $X \sim b(10, 0.01)$. Then, $P(X \geq 1) = 1 - P(X = 0) = 1 - 0.9044 = 0.0956$ is the probability that the process will be stopped.

4.5 If X denotes the number of defective joints and $X \sim b(5, 0.15)$,
  (a) $P(X = 0) = C(5, 0)(0.15)^0(0.85)^5 \approx 0.4437$
  (b) $P(X = 1) = C(5, 1)(0.15)^1(0.85)^4 \approx 0.3915$
  (c) $P(X = 2) = C(5, 2)(0.15)^2(0.85)^3 \approx 0.1382$
  (d) $P(X \geq 2) = 1 - [P(X = 0) + P(X = 1)] = 0.1648$

4.7 Let X denote the number of nonconforming cases in a random sample of 200 cases. If the shipment is very large, $X \approx b(200, 0.05)$ and $P(X = 6) \approx C(200, 6)(0.05)^6(0.95)^{194} \approx 0.0614$.

4.9 In each case, let X denote the number of defective nails in the sample.
  (a) Assuming $X \sim b(400, 0.015)$,
$$\begin{aligned}P(X < 3) &= P(X = 0) + P(X = 1) + P(X = 2) \\ &= C(400, 0)(0.015)^0(0.985)^{400} + C(400, 1)(0.015)^1(0.985)^{399} \\ &\quad + C(400, 2)(0.015)^2(0.985)^{398} \\ &= (0.985)^{400} + (400)(0.015)^1(0.985)^{399} \\ &\quad + (200)(399)(0.015)^2(0.985)^{398} \approx 0.0606.\end{aligned}$$
  (b) Assuming $X \sim b(64, 0.015)$,
$$\begin{aligned}P(X \leq 3) &= P(X = 0) + P(X = 1) + P(X = 2) + P(X = 3) \\ &= C(64, 0)(0.015)^0(0.985)^{64} + C(64, 1)(0.015)^1(0.985)^{63} \\ &\quad + C(64, 2)(0.015)^2(0.985)^{62} + C(64, 3)(0.015)^3(0.985)^{61} \\ &\approx 0.3801 + 0.3705 + 0.1777 + 0.0559 = 0.9842.\end{aligned}$$
  Therefore, $P(X > 3) = 1 - P(X \leq 3) \approx 1 - 0.9842 = 0.0158$.

4.11 Let X denote the number of damaged cases in a random sample of 12 cases. For a large shipment, X has an approximate binomial distribution with probability of success the proportion of damaged cases in the shipment. Thus, we assume that $X \sim b(12, 0.30)$. Using Appendix B, $P(X \leq 5) \approx 0.8822$.

4.13 Let X denote the number of defective switches in the random sample.

  (a) Assuming $X \sim b(500, 0.001)$, $\mu = (500)(0.001) = 0.5$,
  $\sigma^2 = (500)(0.001)(0.999) = 0.4995$, and $\sigma = \sqrt{0.4995} \approx 0.71$.

  (b) If the lot is really 0.1% defective and 4 defective switches are found in the random sample of 500 switches, an outcome that is more than 4 standard deviations above the mean has been realized $[0.5 + 4(0.71) = 3.34]$. This is so unusual that we have strong evidence that the supplier has not complied with the guarantee.

4.15 **(a)** Let X denote the number of customers in a sample of 25 who purchase regular unleaded or super regular gasoline. For the conditions given, we assume that $X \sim b(25, 0.90)$. Using Appendix B, $P(X \geq 10) = 1 - P(X \leq 9) \approx 1 - 0.0000 = 1$.

  (b) If fewer than 3 customers purchase super unleaded, a least 23 customers purchase regular unleaded or super regular. This occurs with probability $P(X \geq 23) = 1 - P(X \leq 22) = 1 - 0.4629 = 0.5371$.

4.17 If the machine is in control, we can assume a sequence of independent Bernoulli trials is observed.

  (a) A sequence of 9 good items followed by a defective item must occur. This happens with probability $(0.92)^9(0.08) \approx 0.0378$.

  Note: A random variable of this type is said to have a *geometric distribution*. For such a variable, a sequence of independent Bernoulli trials, each with probability of success θ, occurs until the first success is observed. Letting X denote the trial on which that success occurs, $f(x) = P(X = x) = (1 - \theta)^{x-1}\theta$; $x = 1, 2, 3, 4, 5, \ldots$.

  (b) There are $C(11, 3) = 165$ possible sequences of 3 defective items and 8 good items. Each occurs with probability $(0.08)^3(0.92)^8 \approx 0.0003$. The probability that the 12th item is defective is 0.08, regardless of the status of the preceding 11 items. Therefore, the probability that the 4th defective item found is the 12th item inspected is $165(0.08)^3(0.92)^8(0.08) \approx 0.0035$.

  Note: A random variable of this type is said to have a *negative binomial distribution*. For such a variable, a sequence of independent Bernoulli trials, each with probability of success θ, occurs until the rth success is observed. If X denotes the trial on which that success occurs, $f(x) = P(X = x) = C(x - 1, r - 1)\theta^r(1 - \theta)^{x-r}$; $x = r, r + 1, r + 2, \ldots$.

4.19 **(a)** If $X \sim b(100, 0.02)$,
$$\begin{aligned} P(X \geq 7) &= 1 - P(X \leq 6) \\ &= 1 - [(0.98)^{100} + C(100, 1)(0.02)(0.98)^{99} \\ &\quad + \ldots + C(100, 6)(0.02)^6(0.98)^{94}] \\ &\approx 1 - 0.9959 = 0.0041. \end{aligned}$$

(b) Since observing 7 or more modules in 100 that require service when $\theta = 0.02$ is so rare, we have strong evidence that more than 2% of the control modules will require warranty service.

Section 4.2.2

4.21 Since $\bar{p} = \dfrac{9 + 15 + \ldots + 14 + 11}{40(14)} = \dfrac{1}{4}$ and $n\bar{p} = 40(\frac{1}{4}) = 10$,

CL $= n\bar{p} = 10$. Thus, UCL $= 10 + 3\sqrt{40(1/4)(3/4)} \approx 10 + 8.2 = 18.2$ and

LCL $= 10 - 3\sqrt{40(1/4)(3/4)} \approx 10 - 8.2 = 1.8$. All points on the following control chart are between the control limits, and no unusual patterns are present. We have insufficient evidence to reject H_0: "The samples were obtained from a process with constant probability of success."

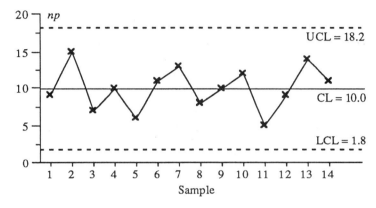

4.23 Since $n = 25$ and $\theta = 0.06$, CL $= 25(0.06) = 1.50$. Furthermore, $3\sqrt{25(0.06)(0.94)} \approx 3.56$, so UCL $\approx 1.50 + 3.56 = 5.06$. Since $1.50 - 3.56 < 0$, we let LCL $= 0.00$.

4.25 $\bar{p} = \dfrac{1 + 2 + \ldots + 1 + 0}{25(20)} = \dfrac{13}{250}$, $3\sqrt{25(13/250)(237/250)} \approx 3.33$, and

$n\bar{p} = 25(13/250) = 1.30$. So, UCL $\approx 1.30 + 3.33 = 4.63$ and CL $= 1.30$. Since $1.30 - 3.33 < 0$, LCL $= 0.00$.

Using JMP®, the following control chart was obtained. Since $n\bar{p}$ is the average of the 20 observed x-values, JMP® uses Avg to denote the centerline. Also, the numeral 1 is placed by each point beyond the control limits. In this case, the point for sample 15 is such a point. Thus, we reject H_0:"The samples were obtained from a process with constant probability of success." Conditions at the time sample 15 was obtained should be studied carefully to see if a reason for this unusual occurrence can be identified.

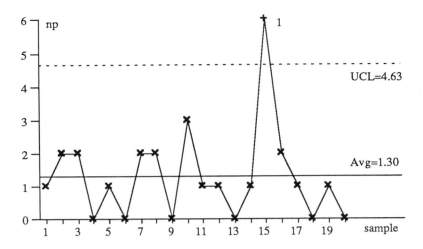

4.27 **(a)** $\bar{p} = \dfrac{272 + 274 + \ldots + 176 + 188}{800(24)} = \dfrac{5{,}167}{19{,}200}$

$CL = n\bar{p} = 800(5{,}167/19{,}200) = 5{,}167/24 \approx 215.3$

$3\sqrt{800(5{,}167/19{,}200)(14{,}033/19{,}200)} \approx 37.6$

$UCL \approx 215.3 + 37.6 = 252.9$ and $LCL \approx 215.3 - 37.6 = 177.7$

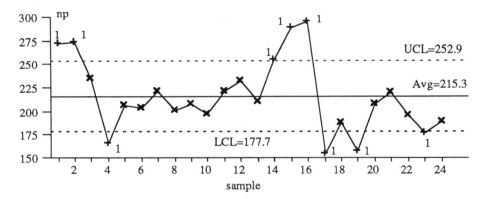

Using JMP®, the preceding control chart was obtained. Since $n\bar{p}$ is the average of the 24 observed x-values, JMP® uses *Avg* to denote the centerline. Also, the numeral 1 is placed beside a point that falls outside the control limits to indicate that an unusually large or small value has been observed.

The counts for samples 1, 2, 14, 15, and 16 exceed the upper control limit, indicating that the numbers of loose nuts in those samples are unusually large (when compared with the other samples). If the cause(s) of this can be identified, appropriate action may result in process improvement.

36

The counts for samples 4, 17, 19, and 23 are below the lower control limit, indicating times during which the process is operating at a much better quality level. If the reasons for this improvement can be found and applied to the process, perhaps the overall average number of loose nuts can be lowered.

(b) In (a), we noted that the counts for samples 1, 2, 14, 15, and 16 are unusually large (when compared with the other samples). Since each of those samples was obtained when a new operator was assembling the bumpers, a cause for this may have been identified. Perhaps those operators need additional training.

Section 4.3

4.29 Let X and Y denote the number of typographical errors in 10 and 30 pages, respectively. Assuming $X \sim \text{Poi}(1)$, $Y \sim \text{Poi}(3)$. Using Appendix C, $P(Y = 0) = 0.0498$. (*Note:* For $Y \sim \text{Poi}(3)$, $P(Y = 0) = e^{-3} \approx 0.049787$.)

4.31 Let X denote the number of misprints on a page. Assuming a Poisson distribution with $\lambda = 400/400 = 1$ and using Appendix C, $P(X \geq 3) = 1 - P(X \leq 2) = 1 - 0.9197 = 0.0803$.

4.33 For $X \sim \text{Poi}(20)$, $P(X \geq 30) = 1 - P(X \leq 29) = 1 - 0.9782 = 0.0218$, from Appendix C.

4.35 (a) Let X denote the number of calls received in 10 minutes. Assuming X has a Poisson distribution with $\lambda = 6/6 = 1$, and using Appendix C, $P(X = 0) = 0.3679$.

(b) Let X denote the number of calls received in 4 hours. Assuming X has a Poisson distribution with $\lambda = 4(6) = 24$, and using Appendix C, $P(X \leq 15) = 0.0344$.

(c) Let Y denote the number of calls received during three ten-minute breaks. Assuming independence, and using the result in (a), $P(Y = 0) = [P(X = 0)]^3 = (0.3679)^3 \approx 0.0498$.

4.37 The average number of defects per sheet is $9(4/6) = 6$. Using Appendix C with $X \sim \text{Poi}(6)$, $P(X \geq 2) = 1 - P(X \leq 1) = 1 - 0.0174 = 0.9826$.

4.39 (a) The mean of $X \sim b(4, 0.03)$ is $\mu = 4(0.03) = 0.12$. Using a Poisson approximation with $Y \sim \text{Poi}(0.12)$, $P(X = 0) \approx P(Y = 0) = e^{-0.12} \approx 0.8869$. From Problem 4.8(a), $P(X = 0) = 0.8853$. Even though the rule of thumb is violated, the approximation is good.

(b) The mean of $X \sim b(10, 0.03)$ is $\mu = 10(0.03) = 0.30$. Using a Poisson approximation with $Y \sim \text{Poi}(0.30)$, $P(X < 2) = P(X \leq 1) \approx P(Y \leq 1) = 0.9631$ from Appendix C. In Problem 4.8(b), we found $P(X < 2) = 0.9655$. Even though the rule of thumb is violated, the approximation is good.

(c) The mean of $X \sim b(5, 0.03)$ is $\mu = 5(0.03) = 0.15$. Using a Poisson approximation with $Y \sim$ Poi(0.15), $P(X \geq 1) = 1 - P(X = 0) \approx 1 - P(Y = 0) = 1 - 0.8607 = 0.1393$ from Appendix C. In Problem 4.8(c), we found $P(X \geq 1) = 0.1413$. Even though the rule of thumb is violated, the approximation is good.

(d) The mean of $X \sim b(6, 0.03)$ is $\mu = 6(0.03) = 0.18$. Using a Poisson approximation with $Y \sim$ Poi(0.18), $P(X = 1) \approx P(Y = 1) = P(Y \leq 1) - P(Y = 0) = 0.9856 - 0.8353 = 0.1503$ from Appendix C. In Problem 4.8(d), we found $P(X = 1) = 0.1545$. Even though the rule of thumb is violated, the approximation is fair.

4.41 Let X denote the number of defective castings in a sample of 100. If the process is stable, we may assume $X \sim b(100, 0.005)$.

(a) $P(X = 1) = 100(0.005)^1 (0.995)^{99} \approx 0.3044$

(b) $P(X \leq 1) = (0.995)^{100} + 100(0.005)^1 (0.995)^{99} \approx 0.9102$

(c) $P(X \geq 1) = 1 - P(X = 0) = 1 - (0.995)^{100} \approx 0.3942$

(d) $P(X \leq 2) = (0.995)^{100} + 100(0.005)^1 (0.995)^{99} + 50(99)(0.005)^2 (0.995)^{98}$
≈ 0.9859

4.43 (a)

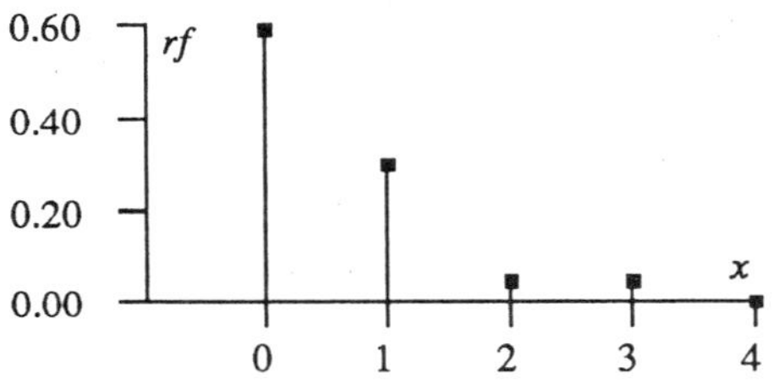

(b) $\bar{x} = [0(12) + 1(6) + 2(1) + 3(1)]/20 = 11/20 = 0.55$

$s^2 = \dfrac{(0-0.55)^2(12)+(1-0.55)^2(6)+(2-0.55)^2(1)+(3-0.55)^2(1)}{19} \approx 0.68$

The values are not greatly different.

(c)

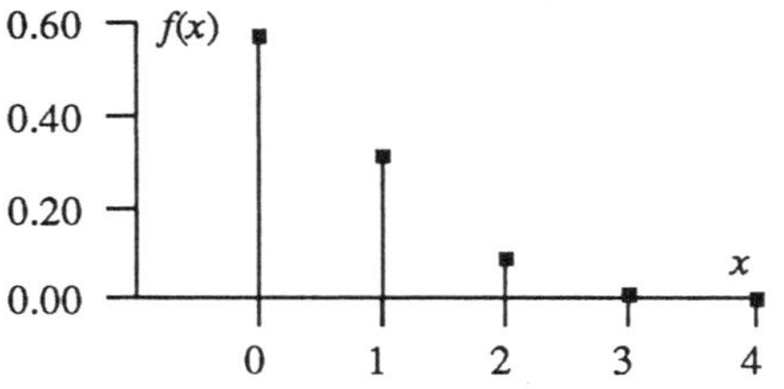

The two graphs are quite similar.

(d) The sample mean and variance do not differ greatly. Also, the relative frequency distribution and the distribution of a Poisson random variable with mean 0.55 are quite similar. Based on these facts, the Poisson assumption is reasonable.

4.45 Let C denote the number of defects in a bar. For the 15 bars, $\bar{c} = 12.4$ and $s^2 \approx 22.54$. Since the sample variance is almost twice the sample mean, a Poisson assumption is questionable. Noting that $s \approx 4.7$, we shall use control limits of UCL $= \bar{c} + 3s = 12.4 + 3(4.7) = 26.5$ and CL $= \bar{c} = 12.4$. Since $\bar{c} - 3s < 0$, let LCL $= 0.0$.

A control chart based on these limits follows. The chart gives no strong evidence in favor of rejecting H_0: "The counts were obtained from a process with a constant average count."

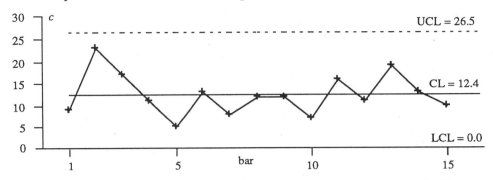

4.47 For $\lambda = 1.3$, UCL $= 1.3 + 3\sqrt{1.3} \approx 4.7$ and CL $= 1.3$. Since $1.3 - 3\sqrt{1.3} < 0$, LCL $= 0$. Using these limits and plotting the given points produces the following chart. The number of flaws for bolt 19 is beyond the upper control limit. This indicates that the standard given is not being met.

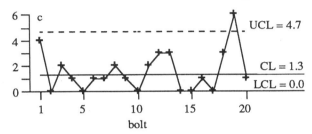

4.49 The plotted points are the same in both graphs. The control limits and hypotheses for the two graph differ. In Problem 4.47, we were able to reject H_0: "The average number of flaws per bolt is 1.3." In Problem 4.48, the hypothesis H_0:"The bolts were obtained from a process with a constant average number of flaws per bolt." was rejected.

4.51 For $\lambda = 5$, UCL $= 5 + 3\sqrt{5} \approx 11.7$ and CL $= 5.0$. Since $5 - 3\sqrt{5} < 0$, LCL $= 0$. Using these limits and plotting the given points produces the following chart. The process may be meeting the stated standard of 5 defects per 100 feet.

39

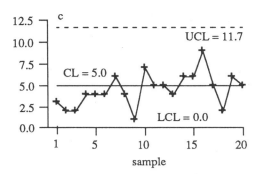

4.53 Since $\bar{c} \approx 51.5$ and $s^2 \approx 440.16$ are quite different, a Poisson assumption is not valid. Using $\bar{c} \pm 3s$, the control limits are: LCL ≈ 0.0, CL ≈ 51.5, and UCL ≈ 114.5. The control chart, given below, does not provide evidence to reject H_0: "The counts were obtained from a process with a constant average count."

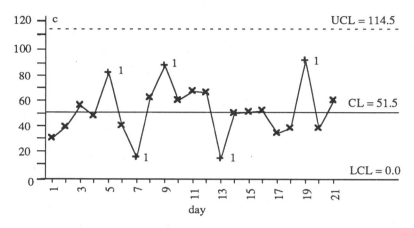

4.55

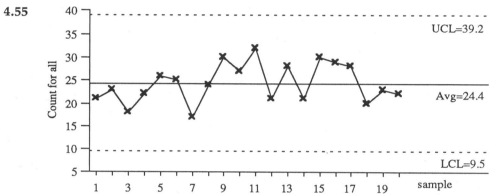

40

The preceding JMP control chart is a chart for the total of all types of blemishes. That chart contains no unusual trends or runs, and no points are outside the control limits. [*Note*: Since $\bar{c}$ is the average of the 20 counts, JMP uses *Avg* to denote the centerline.]

The following c chart for pits indicates that an unusual number of pits occurred in the 11th sample. JMP indicates that by affixing 1 beside the plotted point. A study of the process at the time may enable the investigator to identify a reason for such an occurrence and improve the process by avoiding a similar situation in the future.

JMP also affixed a 2 at the 20th point in the chart for pits. This indicates that the 20th point is the ninth of 9 points in a row that are on the same side of the center line (see Section 8.2.2). In this case, an unusual reduction in the number of pits has been observed. A cause should be sought in an attempt to continue the improvement.

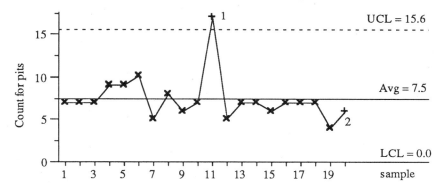

Section 4.4

4.57 Let X denote the number of defective casings in the sample.
$P(X = 1) = C(8, 1)C(992, 2)/C(1000, 3) = (8)(491,536)/166,167,000 \approx 0.0237$

4.59 If X denotes the number of defective items in the sample, $X \sim h(3, 30, 5)$.
(a) $P(X = 0) = C(5, 0)C(25, 3)/C(30, 3) = (1)(2,300)/4,060 \approx 0.5665$
(b) $P(X = 3) = C(5, 3)C(25, 0)/C(30, 3) = (10)(1)/4,060 = 1/406 \approx 0.0025$
(c) $P(X = 1) = C(5, 1)C(25, 2)/C(30, 3) = (5)(300)/4,060 \approx 0.3695$
(d) $P(X \geq 2) = 1 - [P(X = 0) + P(X = 1)] \approx 1 - [0.5665 + 0.3695] = 0.0640$, from (a) and (c).

4.61 Let X denote the number of calculators in the sample that have defective switches.
(a) $P(X = 2) = (10/250)(9/249) \approx 0.0014$
(b) $P(X = 0) = (240/250)(239/249) \approx 0.9214$
(c) $P(X \geq 1) = 1 - P(X = 0) \approx 1 - 0.9124 = 0.0786$ from (b)

4.63 **(a)** For the following table, probabilities were calculated using the hypergeometric function in EXCEL.

41

D	0	1	2	3	4	5	10	15
p	0.0000	0.0125	0.0250	0.0375	0.0500	0.0625	0.1250	0.1875
p_1	1.0000	1.0000	0.9981	0.9944	0.9890	0.9820	0.9258	0.8423

D	20	25	30	40	50	60	70	80
p	0.2500	0.3125	0.3750	0.5000	0.6250	0.7500	0.8750	1.0000
p_1	0.7411	0.6303	0.5174	0.3077	0.1457	0.0463	0.0054	0.0000

(b) Plotting the points summarized in the preceding table produces the following graph.

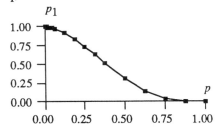

Section 4.5

4.65 **(a)** $P_{acc} = C(50 - 50p, 5)/C(50, 5) + C(50p, 1)C(50 - 50p, 4)/C(50, 5)$ for lot qualities of $p = 0.00, 0.02, \ldots, 1.00$. Using this information with EXCEL produces the following table. [*Note:* $C(n, r) = 0$ when $n < r$.]

p	0.00	0.02	0.04	0.06	0.08
D	0	1	2	3	4
P_{acc}	1.0000	1.0000	0.9918	0.9765	0.9550

p	0.10	0.20	0.30	0.40	0.50
D	5	10	15	20	25
P_{acc}	0.9282	0.7419	0.5239	0.3259	0.1743

(b)

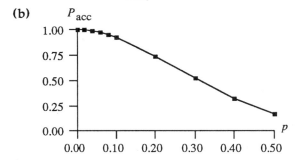

(c) Since $P_{acc} = 1.00$ at $p = 0.02$, $\alpha = 1.00 - 1.00 = 0.00$.

(d) Since $P_{acc} = 0.9765$ at $p = 0.06$, the consumer's risk is $\beta = 0.9765$.

4.67 **(a)** For p the lot quality, $P_{acc} = C(200 - 200p, 5)/C(200, 5)$; $p = 0.000$, $0.005, \ldots, 1.000$. Using this formula with EXCEL produces the following table. [*Note*: $C(n, r) = 0$ when $n < r$.]

p	0.000	0.005	0.010	0.015	0.020	0.030
D	0	1	2	3	4	6
P_{acc}	1.0000	0.9750	0.9505	0.9265	0.9030	0.8574

p	0.040	0.050	0.060	0.080	0.100
D	8	10	12	16	20
P_{acc}	0.8136	0.7717	0.7315	0.6562	0.5872

(b) P_{acc}

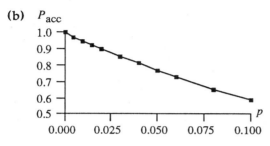

(c) Since $P_{acc} = 0.9505$ at $p = 0.01$, $\alpha = 1 - 0.9505 = 0.0495$.

(d) Since $P_{acc} = 0.8136$ at $p = 0.04$, the consumer's risk is $\beta = 0.8136$.

4.69 **(a)** For p the lot quality, $P_{acc} = C(200 - 200p, 10)/C(200, 10)$; $p = 0.000$, $0.005, \ldots, 1.000$ is used to produce the following table.

p	0.000	0.005	0.010	0.015	0.020	0.030
D	0	1	2	3	4	6
P_{acc}	1.0000	0.9500	0.9023	0.8567	0.8132	0.7321

p	0.040	0.050	0.060	0.080	0.100
D	8	10	12	16	20
P_{acc}	0.6584	0.5915	0.5307	0.4257	0.3398

(b) P_{acc}

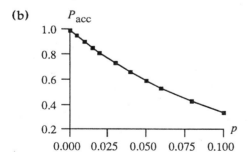

(c) Since $P_{acc} = 0.9023$ at $p = 0.01$, $\alpha = 1 - 0.9023 = 0.0977$.

(d) Since $P_{acc} = 0.6584$ at $p = 0.04$, the consumer's risk is $\beta = 0.6584$.

4.71 Let X denote the number of nonconforming items in the sample. Using $X \approx b(15, p)$, $P_{acc} = P(X = 0) \approx (1 - p)^{15}$. Using this formula, we find the following probabilities.

p	0.000	0.005	0.010	0.015	0.020	0.025
P_{acc}	1.0000	0.9276	0.8601	0.7972	0.7386	0.6840

p	0.030	0.035	0.040	0.050	0.060
P_{acc}	0.6333	0.5860	0.5421	0.4633	0.3953

 (a) Since $P_{acc} \approx 0.8601$ at $p = 0.010$, $\alpha \approx 1 - 0.8601 = 0.1399$.
 (b) Since $P_{acc} \approx 0.4633$ at $p = 0.050$, the consumer's risk is $\beta \approx 0.4633$.
 (c)

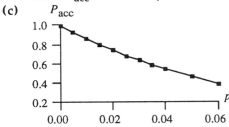

Section 4.5.1

4.73 (a) Using $AOQ = p(45/50)P_{acc}$, we obtain the results tabled below. From the table, we see that AOQL = 0.0578.

p	0.00	0.02	0.04	0.06	0.08	0.10	0.12
P_{acc}	1.0000	0.9000	0.8082	0.7240	0.6470	0.5766	0.5126
AOQ	0.0000	0.0162	0.0291	0.0391	0.0466	0.0519	0.0554

p	0.14	0.16	0.18	0.20	0.22	0.24	0.26
P_{acc}	0.4543	0.4015	0.3537	0.3106	0.2717	0.2369	0.2057
AOQ	0.0572	0.0578	0.0573	0.0559	0.0538	0.0512	0.0481

p	0.28	0.30	0.32	0.34	0.36	0.38
P_{acc}	0.1779	0.1532	0.1313	0.1120	0.0950	0.0802
AOQ	0.0448	0.0414	0.0378	0.0343	0.0308	0.0274

 (b) A plot of the AOQ curve follows. Since AOQL = 0.0578, the point at which the maximum occurs is (0.16, 0.0578).

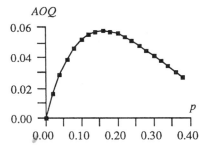

4.75 **(a)** Using $AOQ = p(40/50)P_{acc}$, we obtain the following tabled results. From the table, we see that AOQL = 0.0254.

p	0.00	0.02	0.04	0.06	0.08	0.10	0.12
P_{acc}	1.0000	0.8000	0.6367	0.5041	0.3968	0.3106	0.2415
AOQ	0.0000	0.0128	0.0204	0.0242	0.0254	0.0248	0.0232

p	0.14	0.16	0.18	0.20	0.22	0.24	0.26
P_{acc}	0.1867	0.1432	0.1091	0.0825	0.0619	0.0460	0.0339
AOQ	0.0209	0.0183	0.0157	0.0132	0.0109	0.0088	0.0071

p	0.28	0.30	0.32	0.34	0.36	0.38
P_{acc}	0.0247	0.0179	0.0128	0.0090	0.0063	0.0043
AOQ	0.0055	0.0043	0.0033	0.0025	0.0018	0.0013

(b) A plot of the AOQ curve follows. Since AOQL = 0.0254, the point at which the maximum occurs is (0.08, 0.0254).

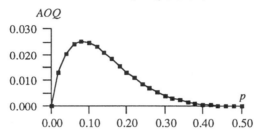

4.77 **(a)** Using $AOQ = p(195/200)P_{acc}$ and EXCEL, the following table results. From the table, we see that AOQL ≈ 0.1554.

p	0.000	0.005	0.010	0.015	0.020	0.025	0.030	0.035
P_{acc}	1.0000	1.0000	0.9995	0.9985	0.9970	0.9951	0.9928	0.9900
AOQ	0.0000	0.0049	0.0097	0.0146	0.0194	0.0243	0.0290	0.0338

p	0.040	0.045	0.050	0.055	0.060	0.065	0.070	0.075
P_{acc}	0.9868	0.9832	0.9792	0.9748	0.9701	0.9650	0.9596	0.9539
AOQ	0.0385	0.0431	0.0477	0.0523	0.0568	0.0612	0.0655	0.0698

p	0.080	0.085	0.090	0.095	0.100	0.105	0.110	0.115
P_{acc}	0.9478	0.9415	0.9348	0.9279	0.9208	0.9134	0.9057	0.8978
AOQ	0.0739	0.0780	0.0820	0.0860	0.0898	0.0935	0.0971	0.1007

p	0.120	0.125	0.130	0.135	0.140	0.145	0.150	0.155
P_{acc}	0.8897	0.8814	0.8729	0.8642	0.8553	0.8463	0.8371	0.8278
AOQ	0.1041	0.1074	0.1106	0.1138	0.1168	0.1196	0.1224	0.1251

p	0.160	0.165	0.170	0.175	0.180	0.185	0.190	0.195
P_{acc}	0.8183	0.8086	0.7989	0.7890	0.7791	0.7690	0.7589	0.7486
AOQ	0.1276	0.1301	0.1324	0.1346	0.1367	0.1387	0.1406	0.1423

p	0.200	0.205	0.210	0.215	0.220	0.225	0.230	0.235
P_{acc}	0.7383	0.7280	0.7175	0.7071	0.6965	0.6860	0.6754	0.6648
AOQ	0.1440	0.1455	0.1469	0.1482	0.1494	0.1505	0.1515	0.1523

p	0.240	0.245	0.250	0.255	0.260	0.265	0.270	0.275
P_{acc}	0.6541	0.6435	0.6328	0.6222	0.6115	0.6009	0.5903	0.5797
AOQ	0.1531	0.1537	0.1543	0.1547	0.1550	0.1553	0.1554	0.1554

p	0.280	0.285	0.290	0.295
P_{acc}	0.5691	0.5586	0.5481	0.5376
AOQ	0.1554	0.1552	0.1550	0.1546

(b) The ordinate of the point (0.275, 0.1554) on the following AOQ curve is AOQL $\approx$ 0.1554.

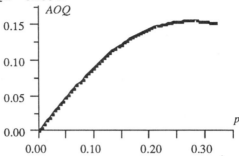

4.79 Using $AOQ \approx p(P_{acc})$ and EXCEL, the following table results. From the table, we see that AOQL $\approx$ 0.0237.

p	0.000	0.005	0.010	0.015	0.020	0.025	0.030
P_{acc}	1.0000	0.9276	0.8601	0.7972	0.7386	0.6840	0.6333
AOQ	0.0000	0.0046	0.0086	0.0120	0.0148	0.0171	0.0190

P	0.035	0.040	0.045	0.050	0.055	0.060	0.065
P_{acc}	0.5860	0.5421	0.5012	0.4633	0.4280	0.3953	0.3649
AOQ	0.0205	0.0217	0.0226	0.0232	0.0235	0.0237	0.0237

P	0.070	0.075	0.080	0.085	0.090	0.095
P_{acc}	0.3367	0.3105	0.2863	0.2638	0.2430	0.2237
AOQ	0.0236	0.0233	0.0229	0.0224	0.0219	0.0213

Supplementary Problems

4.81 Let X denote the number of defective units in the sample. Assuming $X \sim b(15, 0.30)$, and using Appendix B,
   (a) $P(X = 5) = P(X \le 5) - P(X \le 4) = 0.7216 - 0.5155 = 0.2061$
   (b) $P(X > 5) = 1 - P(X \le 5) = 1 - 0.7216 = 0.2784$
   (c) $P(X = 0) = 0.0047$

4.83 Let X denote the number of defective items in the sample. Assuming random selection, $X \sim h(5, 50, 10)$. Thus, $\mu = 5(10/50) = 1$, $\sigma^2 = 5(10/50)(40/50)(45/49) = 36/49 \approx 0.7347$, and $\sigma = \sqrt{36/49} = 6/7 \approx 0.86$.

4.85 The probability that a randomly selected compact disc is returned and repaired is $(0.10)(0.70) = 0.07$, by the multiplication rule. If X denotes the number of compact discs in a random sample of size 30 that are returned and repaired, then $X \sim b(30, 0.07)$. Thus,

$$
\begin{aligned}
P(X \le 1) &= P(X = 0) + P(X = 1) \\
&= (0.93)^{30} + C(30, 1)(0.07)(0.93)^{29} \\
&\approx 0.1134 + 0.2560 = 0.3694.
\end{aligned}
$$

4.87 Let p denote the probability that a solder joint is defective. Since a randomly selected circuit board needs no repair when each of its solder joints is nondefective, and 98% of all boards need no repair, $0.98 = (1 - p)^{20}$. Thus, $1 - p = (0.98)^{0.05} \approx 0.999$ and $p \approx 0.001$.

4.89 (a) For a random drawing, $X \sim b(41, 0.50)$. Thus,

$$
\begin{aligned}
P(X \ge 40) &= P(X = 40) + P(X = 41) \\
&= C(41, 40)(0.50)^{40}(0.50) + C(41, 41)(0.50)^{41}(0.50)^0 \\
&= 42(0.50)^{41} \approx 1.9 \times 10^{-11} \approx 0.
\end{aligned}
$$

(b) Yes. The probability in (a) is so small, that it is unreasonable to believe the clerk determined the position by random drawing.

4.101 Assuming the two variables are independent with $X \sim b(10, 0.03)$ and $Y \sim b(10, 0.03)$, and using Appendix B,

$$
\begin{aligned}
P(\text{process unchanged}) &= P(X = 0) + P(X = 1)P(Y = 0) \\
&= 0.7374 + (0.9655 - 0.7374)(0.7374) \approx 0.9056.
\end{aligned}
$$

Thus, $P(\text{process adjusted}) = 1 - 0.9056 = 0.0944$.

CHAPTER 5
CONTINUOUS PROBABILITY DISTRIBUTIONS

Section 5.1

5.1 Using Appendix E,
   (a) $P(Z \ge 2.12) = 1 - \Phi(2.12) = \Phi(-2.12) = 0.01700$
   (b) $P(Z \le 1.85) = \Phi(1.85) = 0.96784$
   (c) $P(Z > 0.71) = 1 - \Phi(0.71) = \Phi(-0.71) = 0.23885$
   (d) $P(0.50 < Z < 1.38) = \Phi(1.38) - \Phi(0.50) = 0.91621 - 0.69146 = 0.22475$
   (e) $P(-1.49 < Z < 1.67) = \Phi(1.67) - \Phi(-1.49) = 0.95254 - 0.06811 = 0.88443$
   (f) $P(Z > -0.43) = 1 - \Phi(-0.43) = \Phi(0.43) = 0.66640$

5.3 Since $1 - 0.2843 = 0.7157$ and $\Phi(0.57) \approx 0.7157$, $P(Z \ge 0.57) \approx 0.2843$. Therefore, $c \approx 0.57$.

5.5 When $\mu = 85$, $\sigma = \sqrt{36} = 6$, and x is an observed value of X, the standardized value of x is $z = (x - 85)/6$.
   (a) If $x = 75$, $z = (75 - 85)/6 \approx -1.67$.
   (b) If $x = 92$, $z = (92 - 85)/6 \approx 1.17$.
   (c) If $z = 1.31$, $x = 6(1.31) + 85 = 92.86$.
   (d) If $z = -2.4$, $x = 6(-2.4) + 85 = 70.6$.

5.7 Since $X \sim N(2, 4)$, $Z = (X - 2)/2$ has a standard normal distribution. Using Appendix E,

(a) $P(-8 < X < 0) = P[(-8 - 2)/2) < Z < (0 - 2)/2]$
$= P(-5 < Z < -1) \approx \Phi(-1) = 0.15866$

(b) $P(4 < X \le 8) = P[(4 - 2)/2) < Z \le (8 - 2)/2]$
$= P(1 < Z < 3)$
$= \Phi(3) - \Phi(1) = 0.99865 - 0.84134 = 0.15731$

(c) $P(X < -3) = P[Z < (-3 - 2)/2] = P(Z < -2.5) = \Phi(-2.5) = 0.00621$

5.9 Since $X \sim N(80, 16)$, $Z = (X - 80)/4$ has a standard normal distribution. Using Appendix E, $P(74 < X < 82) = P[(74 - 80)/4 < Z < (82 - 80)/4]$
$= P(-1.50 < Z < 0.50)$
$= \Phi(0.50) - \Phi(-1.50)$
$= 0.69146 - 0.06681 = 0.62465.$

5.11 Since $X \sim N(5.35, 0.2025)$, $Z = (X - 5.35)/0.45$ has a standard normal distribution. Using Appendix E,

(a) $P(X > 5.80) = P[Z > (5.80 - 5.35)/0.45] = P(Z > 1) = \Phi(-1) = 0.15866$

(b) $P(X < 4.60) = P[Z < (4.60 - 5.35)/0.45]$
$\approx P(Z < -1.67)$
$= \Phi(-1.67) = 0.04746$

(c) $P(5.50 < X < 5.72) = P[(5.50 - 5.35)/0.45 < Z < (5.72 - 5.35)/0.45] \approx$
$P(0.33 < Z < 0.82) = \Phi(0.82) - \Phi(0.33) = 0.79389 - 0.62930 = 0.16459$

5.13 Let X denote the tension reading. Since $X \sim N(270, 40^2)$, $Z = (X - 270)/40$ has a standard normal distribution. Using Appendix E, $P(X < 200) = P[Z < (200 - 270)/40] = P(Z < -1.75) = 0.04006$. So, about 4% of the display tubes fail to meet the specification.

5.15 Let X denote the width of a slot. Since $X \sim N(0.900, 0.003^2)$, the standardized values of 0.895 and 0.905 are $(0.895 - 0.900)/0.003 \approx -1.67$ and $(0.905 - 0.900)/0.003 \approx 1.67$. From Appendix E, $P(0.895 \le X \le 0.905) \approx \Phi(1.67) - \Phi(-1.67) = 0.95254 - 0.04746 = 0.90508$. The probability that a slot fails to meet the specifications is $1 - 0.90508 = 0.09492$. So, about 9.5% of the forgings fail to meet the specifications.

5.17 If $P(-z \le Z \le z) = 0.99500$, $z \approx 2.81$. Thus, $\Delta \approx 2.81\sigma = (2.81)\sqrt{0.04} = (2.81)(0.20) = 0.5620.$

5.19 The 99.5th quantile of the standard normal distribution is approximately 2.576. Thus, 5.85 is approximately 2.576 standard deviations above the mean. Since $\sigma = 0.38$, $\mu = 5.85 - (2.576)(0.38) \approx 4.87$.

Section 5.1.1

5.21

i	1	2	3	4	5	6	7	8	9	10
$x_{(i)}$	-1	-1	1	2	3	3	4	4	5	5
$p_{(i)}$	0.09	0.18	0.27	0.36	0.45	0.55	0.64	0.73	0.82	0.91
z_i	-1.34	-0.92	-0.61	-0.36	-0.13	0.13	0.36	0.61	0.92	1.34

The points in the normal quantile plot exhibit a linear pattern. A box plot (not shown) reveals no outliers or suspect outliers. An assumption of normality seems reasonable.

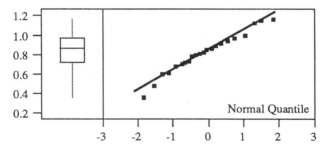

5.23

i	1	2	3	4	5	6	7	8	9	10
$x_{(i)}$	0.36	0.48	0.60	0.61	0.68	0.68	0.71	0.72	0.73	0.79
$p_{(i)}$	0.03	0.06	0.10	0.13	0.16	0.19	0.23	0.26	0.29	0.32
z_i	-1.88	-1.55	-1.28	-1.13	-0.99	-0.88	-0.74	-0.64	-0.55	-0.47

i	11	12	13	14	15	16	17	18	19	20
$x_{(i)}$	0.80	0.81	0.81	0.82	0.85	0.87	0.87	0.89	0.92	0.92
$p_{(i)}$	0.35	0.39	0.42	0.45	0.48	0.52	0.55	0.58	0.61	0.65
z_i	-0.39	-0.28	-0.20	-0.13	-0.05	0.05	0.13	0.20	0.28	0.39

i	21	22	23	24	25	26	27	28	29	30
$x_{(i)}$	0.92	0.94	0.97	0.97	0.97	1.00	1.00	1.13	1.16	1.17
$p_{(i)}$	0.68	0.71	0.74	0.77	0.81	0.84	0.87	0.90	0.94	0.97
z_i	0.47	0.55	0.64	0.74	0.88	0.99	1.13	1.28	1.55	1.88

The normal quantile plot, obtained using JMP and the data in the preceding table, is reasonably linear. No outliers or suspect outliers are indicated on the box plot. An assumption of normality seems reasonable.

Section 5.1.2

5.25 (a) The histogram in the solution to Problem 2.7 is not bell-shaped. This leads us to suspect the data were not obtained from a population associated with a normally distributed random variable. Using MYSTAT to calculate and plot the points, the following normal quantile plot was obtained. The large number of points with the same ordinate may be due to a measurement system problem.

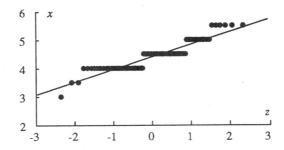

A JMP normal quantile plot follows. To prevent the user from obtaining a graph like the one we constructed with the help of MYSTAT, duplicate values are ignored. Thus, JMP prepared a normal quantile plot based on the six distinct values in the data table. Notice the S shape to the graph. If the data were obtained from a single population, we have reason to believe that population is nonnormal.

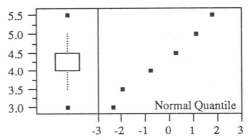

The box plot included in the preceding JMP display indicates the presence of extreme outliers in the data. There are 6 observations of 5.5, two in each of samples 10, 12, and 15. Sample 17 contains the only observation of 3.0. Since the data were obtained in samples of size 5 over a two-week time period, these outliers are probably telling us that more than one population was sampled. Control chart procedures discussed in Chapter 8 can be used to confirm this suspicion.

(b) The normal probability model does not appear to be a good one for estimation of the percent of measurements in the population for the two-week period that exceed the upper specification.

5.27 (a) JMP output that includes a histogram, box plot, and normal quantile plot follows. The box plot indicates the presence of two outliers (8.4 in Sample 9 and 9.6 in Sample 11). Those are consistent with a distribution with a longer upper tail. The distinct, upward curvature in the normal quantile plot indicates such a distribution.

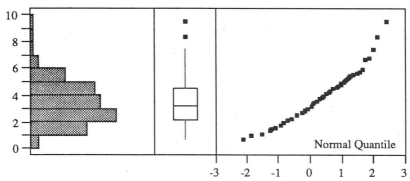

(b) $\bar{x} \approx 3.36$; $s \approx 1.63$

(c) The normal probability model does not appear to be a good one for the population associated with the circuit boards produced during the time of the study.

Section 5.1.3

5.29 Let X denote the number of defective stoves in a random sample of 500 stoves. Assuming $X \sim b(500, 0.05)$, $P(X < 20) = P(X \leq 19.5) \approx P[Z \leq (19.5 - 25)/\sqrt{23.75}] \approx P(Z \leq -1.13) = 0.12924$ from Appendix E.

(*Note*: Using the binomial distribution, $P(X < 20) = P(X \leq 19) \approx 0.12723$.)

5.31 Assuming $Y \sim b(160, 0.05)$, $\mu_Y = 8$ and $\sigma_Y = \sqrt{7.6} \approx 2.76$.
  (a) For $X \sim N(8, 7.6)$, $P(Y \leq 3) \approx P(X \leq 3.5) \approx P[Z \leq (3.5 - 8)/2.76] \approx P(Z \leq -1.63) \approx 0.05155$.
  (b) The normal approximation overestimates the binomial probability more than the Poisson approximation.

5.33 Assuming $X \sim b(100, 0.10)$, $E[X] = 10$ and $Var(X) = 9$.
  (a) X has a hypergeometric distribution, but a binomial model should provide a good approximation.
  (b) If $X \sim b(100, 0.10)$, $P(X \leq 9) = 0.45129$.
  (c) For $Y \sim N(10, 9)$, $P(X \leq 9) \approx P(Y \leq 9.5) = P[Z \leq (9.5 - 10)/3] \approx P(Z \leq -0.17) = 0.43251$.
  (d) The normal probability is a fair approximation to the binomial probability.

Section 5.1.4

5.35 $W \sim N(44, 25)$, so $P(W > 50) = P[Z > (50 - 44)/5] = P(Z > 1.20) = 0.11507$.

5.37 For $Y = X_1 + X_2$, $\mu_Y = 0.1065 + 0.0827 = 0.1892$ and $\sigma_Y = [0.00035^2 + 0.002^2]^{0.5} \approx 0.00203$. Since the standardized values for 0.195 and 0.205 are $(0.195 - 0.1892)/0.00203 \approx 2.86$ and $(0.205 - 0.1892)/0.00203 \approx 7.78$, respectively, $P(Y < 0.195) + P(Y > 0.205) \approx P(Z < 2.86) = 0.99788$. Thus, about 99.8% of the pairs will not meet specifications.

5.39 Let X and Y denote the outside and inside diameter, respectively. If specifications are set at the natural tolerances, $\mu_X = 0.250$, $\sigma_X = 0.005/3$, $\mu_Y = 0.255$, and $\sigma_Y = 0.005/3$. Letting $W = Y - X$, $\mu_W = \mu_Y - \mu_X = 0.005$ and $\sigma_W = \sqrt{2(0.005/3)^2} \approx 0.00236$. Since W is normally distributed, $P(W \leq 0.001) = P[Z \leq (0.001 - 0.005)/0.00236] \approx P(Z \leq -1.69) = 0.04551$. So, the clearances of about 4.6% of the assemblies are at most 0.001 inch.

5.41 Let S denote the outside diameter of a shaft and B denote the inside diameter of a bushing. If $X = B - S$ denotes the clearance, $\mu_X = \mu_B - \mu_S = 1.06 - 1.05 = 0.01$ inch and $\sigma_X^2 = \sigma_B^2 + \sigma_S^2 = 0.001^2 + \sigma_S^2$. Since X is normally distributed, $P(X \leq 0) = 0.01$ implies that the standardized value of 0 is -2.326. So, $-2.326 = (0 - \mu_X)/\sigma_X$, which implies that $\sigma_X^2 = \mu_X^2/2.326^2 = 0.01^2/2.326^2 \approx 0.00002$. Therefore, $0.001^2 + \sigma_S^2 = 0.00002$, and the maximum allowable value of σ_S is $\sqrt{0.00002 - 0.001^2} \approx 0.004$.

Section 5.2

5.43 The standard deviation of X is $\sigma = \sqrt{12} = 2\sqrt{3}$, so $\mu \pm 0.40\sigma = \mu \pm 0.80\sqrt{3}$. The area under the probability density curve bounded by $x = \mu - 0.80\sqrt{3}$ and $x = \mu + 0.80\sqrt{3}$ is $[(\mu + 0.80\sqrt{3}) - (\mu - 0.80\sqrt{3})]/(207 - 195) = (1.6\sqrt{3})/12 = 0.40/\sqrt{3} \approx 0.2309$. Therefore, the probability that X is within 0.40 standard deviations of the mean is about 0.2309.

5.45 Let $X_{0.30}$ denote the 30th percentile of the distribution of X. Then, $P(X \leq X_{0.30}) = 0.30$. Since $X \sim U(20, 60)$, $P(X \leq X_{0.30}) = (X_{0.30} - 20)/(60 - 20) = (X_{0.30} - 20)/40$. Therefore, $X_{0.30} = 20 + 40(0.30) = 32$.

5.47 Let X_i denote the thickness of the ith plate, for $i = 1, 2, 3$. Let Y_j denote the thickness of the jth spacer, for $j = 1, 2$. If $X_i \sim U(0.098, 0.102)$ and $Y_j \sim U(0.197, 0.203)$, then $E[X_i] = 0.100$, $Var(X_i) = (0.102 - 0.098)^2/12 = 0.000004/3$, $E[Y_j] = 0.200$, and $Var(Y_j) = (0.203 - 0.197)^2/12 = 0.000003$.

Let $W = X_1 + X_2 + X_3 + Y_1 + Y_2$ denote the thickness of the assembly. $E[W] = 3(0.100) + 2(0.200) = 0.700$, $Var(W) = 3(0.000004/3) + 2(0.000003) = 0.000010$, and $\sigma_W = \sqrt{0.000010}$. Since $3\sigma_W = 3\sqrt{0.000010} \approx 3(0.00316) \approx 0.009$, the expected dimensions for W are 0.700 ± 0.009 inch.

Section 5.3

5.49 Let X denote the life length of a component. Since $\mu = 100$, $X \sim Exp(100)$ and $P(X \geq 400) = e^{-400/100} = e^{-4} \approx 0.0183$.

5.51 For $X \sim Exp(5)$, the 5th percentile is $X_{0.05} = (-5) \times \ln(0.95) \approx 0.2565$. Therefore, a warranty period of about 0.2565 years would suffice. This is about 94 days, so a 90 day warranty would be acceptable.

5.53 **(a)** If $X \sim Exp(2000)$, $P(X > 2500) = e^{-2500/2000} = e^{-1.25} \approx 0.2865$

(b) Using 5.52(a) and $X \sim Exp(2000)$, $P(X \geq 3500 \mid X \geq 1000) =$
$P(X \geq 2500) = e^{-2500/2000} = e^{-1.25} \approx 0.2865$.

5.55 Using Equation (5.26) and $X \sim Exp(20)$, $X_{0.25} = -20 \times \ln(1 - 0.25) =$
$-20 \times \ln(0.75) \approx 5.75$.

5.57 Let W denote the waiting time (in months) to the next failure, and let X
denote the number of failures in one month. Since $X \sim Poi(0.40)$,
$W \sim Exp(2.50)$. So, $P(X \leq 1.50) = 1 - e^{-1.50/2.50} = 1 - e^{-0.60} \approx 0.4512$.

5.59 Let X denote the life length of a circuit. Since $X \sim Exp(7)$, $P(X \geq 8) = e^{-8/7}$
≈ 0.3189. If Y denotes the number of circuits in a random sample of 10
such circuits that function for at least 8 years, then $Y \sim b(10, 0.3189)$.
Therefore, $P(Y \geq 2) = 1 - P(Y \leq 1) = 1 - [(0.6811)^{10} + 10(0.3189)(0.6811)^9] \approx$
$1 - 0.1221 = 0.8779$.

Section 5.4

5.61 **(a)** $e^{-2.25} = P(X > 9) = e^{-(9/\theta)^\beta}$ from Equation (5.30). Since $\beta = 2$,
$e^{-2.25} = e^{-81/\theta^2}$. Thus, $2.25 = 81/\theta^2$. So, $\theta^2 = 81/2.25 = 36$.
Since $\theta > 0$, $\theta = 6$.

(b) Using Equation (5.28) with
$\beta = 2$ and $\theta = 6$, we obtain
$f(x) = (x/18)e^{-x^2/36}; x > 0$.
A graph is given to the
right.

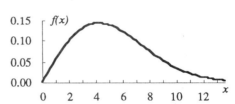

(c) Using Equation (5.30) with $\beta = 2$ and $\theta = 6$, $R(x) = P(X \geq x) = e^{-x^2/36}$;
$x > 0$. From the following graph, we see that $R(5) \approx 0.50$, so the
median life-length is about 5 years. Since $R(9.2) \approx 0.10$, about 90%
of the vehicles last 9.2 years or less.

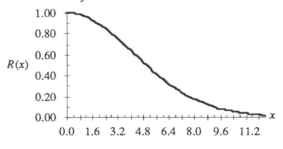

5.63 **(a)**

x	91	122	195	220	261	315	397
$\ln(x)$	4.511	4.804	5.273	5.394	5.565	5.753	5.984
p	0.125	0.250	0.375	0.500	0.625	0.750	0.875
y	-2.013	-1.246	-0.755	-0.367	-0.019	0.327	0.732

Using the $(\ln(x), y)$ pairs summarized in the preceding table, the following MYSTAT graphic is obtained. Notice how well the points fall along a line. This indicates that a Weibull probability model may adequately describe the sampled distribution.

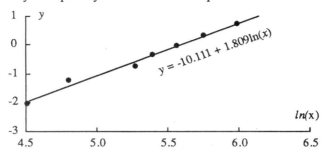

(b) The least squares line has equation $y = -10.111 + 1.809\ln(x)$. Thus, $\beta \approx 1.809$ and $\theta \approx e^{-(-10.111/1.809)} \approx 267.542$.

(c) Approximately 63.2% of the lifetimes are less than 267.542 hours.

5.65 $P(X > 500) = e^{-(500/2000)^{1.5}} = e^{-0.125} \approx 0.8825$

$P(X > 2000) = e^{-(2000/2000)^{1.5}}$ and $P(X > 2500) = e^{-(2500/2000)^{1.5}} = e^{-(1.25)^{1.5}}$.

So, $P(X > 2500 \mid X > 2000) = \dfrac{P(X > 2500 \text{ and } X > 2000)}{P(X > 2000)}$

$= \dfrac{P(X > 2500)}{P(X > 2000)}$

$= e^{-(1.25)^{1.5}} \div e^{-1}$

$= e^{1-(1.25)^{1.5}} \approx 0.6720.$

5.67

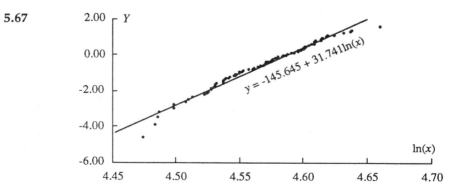

Consider the preceding Weibull plot. Except at the extremes, the points form a reasonably linear pattern. So, a Weibull probability model may be adequate. The points at the extremes should be investigated further.

Section 5.5

5.69 Since $X \sim \Gamma(\alpha, \theta)$ has mean $\mu = \alpha\theta$ and variance $\sigma^2 = \alpha\theta^2$, $\alpha\theta = 8$ and $\alpha\theta^2 = 16$. Solving simultaneously gives $\alpha = 4$ and $\theta = 2$.

(a) Since $\Gamma(4) = 3! = 6$, $f(x) = [1/(\Gamma(4)\times 2^4)]x^3 e^{-x/2} = (\frac{1}{96})x^3 e^{-x/2}$; $x \geq 0$.

(b) Using Chebyshev's rule with $\mu = 8$ and $\sigma = \sqrt{16} = 4$, at least eight-ninths of this population are in the interval $8 \pm 3(4) = [-4, 20]$. But X is never negative, so $[0, 20]$ contains at least eight-ninths of the population.

5.71 (a) Since $\alpha = 2$, $\theta = 0.5$, and $\Gamma(2) = 1$, $f(x) = [1/(0.5)^2]xe^{-2x} = 4xe^{-2x}$; $x \geq 0$.

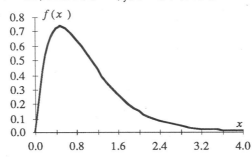

(b) $P(X \leq 1) = 1 - 3e^{-2} \approx 0.5940$

$P(X > 2) = 1 - P(X \leq 2) = 1 - [1 - 5e^{-4}] \approx 0.0916$

5.73 For V the number of crashes per month, $V \sim Poi(1/6)$. So,

(a) $X \sim Exp(6)$ and $P(X \geq 12) = e^{-12/6} = e^{-2} \approx 0.1353$

(b) $Y \sim \Gamma(3, 6)$ and $P(Y \geq 24) = e^{-4}[1 + 4 + (4^2/2!)] = 13e^{-4} \approx 0.2381$

(c) If U denotes the number of crashes in one year, then $U \sim Poi(2)$ and $P(U \geq 3) = 1 - P(U \leq 2) = 1 - 0.6767 = 0.3233$.

5.75 If X denotes the number of forms containing at least one error that arrive during a one-hour period, $X \sim Poi(0.4)$.

(a) From the discussion surrounding Equation (5.45), if T denotes the number of hours until the second form containing an error arrives, then $T \sim \Gamma(2, 2.5)$. So, $P(T \geq 12) = e^{-4.8}[1 + 4.8] \approx 0.0477$.

(b) If Y denotes the number of forms containing at least one error that arrive during the first 4 hours, then $Y \sim Poi(1.6)$. Using Appendix C, $P(Y = 0) = 0.2019$

Chapter 5 Supplementary Problems

5.77 From Example 5.9, the probability that a randomly selected plated bracket meets specifications is approximately 0.945. Letting Y denote the number of brackets in a random sample of size 25 that meet specifications, $Y \sim b(25, 0.945)$. So,

$$P(Y \geq 22) = C(25, 22)(0.945)^{22}(0.055)^3 + C(25, 23)(0.945)^{23}(0.055)^2$$
$$+ C(25, 24)(0.945)^{24}(0.055) + (0.945)^{25}$$
$$\approx 0.95411.$$

5.89   Let $\beta = 2$ and $\theta = 6$. Using the formulas in Problem 5.88, $\Gamma(r + 1) = r\Gamma(r)$, and $\Gamma(0.5) = \sqrt{\pi}$:

$$E[X] = 6\Gamma\left(\tfrac{1}{2} + 1\right) = 6\left(\tfrac{1}{2}\right)\Gamma\left(\tfrac{1}{2}\right) = 3\sqrt{\pi}$$

$$Var(X) = \left(6^2\right)\left[\Gamma\left(\tfrac{2}{2} + 1\right) - \left(\Gamma\left(\tfrac{1}{2} + 1\right)\right)^2\right] = (36)\left[1! - \left((0.5)\sqrt{\pi}\right)^2\right] = 9(4 - \pi)$$

CHAPTER 6
SAMPLING DISTRIBUTIONS AND ESTIMATION

Section 6.1.1

6.1 $\overline{X}$ is normally distributed with mean 1.526 ounces and standard deviation 0.101/5 ounce.

$$
\begin{aligned}
P(1.490 < \overline{X} < 1.572) &= P\left(\tfrac{5(1.490-1.526)}{0.101} < Z < \tfrac{5(1.572-1.526)}{0.101}\right) \\
&\approx P(-1.78 < Z < 2.28) \\
&= \Phi(2.28) - \Phi(-1.78) \\
&= 0.98870 - 0.03754 = 0.95116
\end{aligned}
$$

6.3   The sampled population has mean 10 and variance 100. Therefore, $\overline{X}$ has mean 10 and variance $100/3 \approx 33.3$.

Section 6.1.2

6.5   **(a)** Possible ordered samples are {0, 0}, {0, 1}, {0, 5}, {1, 0}, {1, 1}, {1, 5}, {5, 0}, {5,1}, and {5,5}. Since sampling is with replacement, each occurs with probability $(1/3)(1/3) = 1/9$.

(b) Calculating the mean for each sample in (a), we obtain the following table.

$\overline{x}$	0.0	0.5	1.0	2.5	3.0	5.0
$f(\overline{x})$	⅑	2/9	⅑	2/9	2/9	⅑

The only samples having mean 0.5 are {0, 1} and {1, 0}. Since outcomes are equally likely, $f(0.5) = P\left(\overline{X} = 0.5\right) = 2/9$. The other probabilities are obtained likewise.

(c)

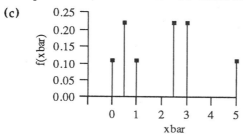

56

(d) $E[\overline{X}] = (0.0)(\tfrac{1}{9})+(0.5)(\tfrac{2}{9})+(1.0)(\tfrac{1}{9})+$
$\qquad\qquad (2.5)(\tfrac{2}{9})+(3.0)(\tfrac{2}{9})+(5.0)(\tfrac{1}{9})$
$\qquad = {}^{18}\!/_{9} = 2.0$

Notice that $E[\overline{X}] = E[X]$.

$Var[\overline{X}] = (0.0)^2(\tfrac{1}{9}) + (0.5)^2(\tfrac{2}{9}) + (1.0)^2(\tfrac{1}{9}) + (2.5)^2(\tfrac{2}{9}) +$
$\qquad\qquad (3.0)^2(\tfrac{2}{9}) + (5.0)^2(\tfrac{1}{9}) - 2^2$
$\qquad = ({}^{57}\!/_9) - 4 = {}^{21}\!/_9 = {}^{7}\!/_3$

Notice that $Var[\overline{X}] = Var[X]/2$.

6.7 The distributions of both are more bell-shaped than that of the sampled population. Neither is very "normal", however. The means are the same. The variance is least for the mean of the distribution derived for the larger sample size. Since the sample sizes are 2 and 3 for problems 6.5 and 6.6, respectively, the results obtained in both problems agree with Equations (6.2) and (6.3).

6.9 $\overline{X}$ is an unbiased estimator of the mean. In this case, $\tilde{X}$ is a biased estimator of μ ($E[\tilde{X}] < \mu$) and has a larger variance than $\overline{X}$. Thus, $\overline{X}$ is preferred when estimating μ.

6.11 **(a)** Using the relative frequencies,
$\qquad \overline{\overline{x}} = 2(0.015) +...+ 11(0.029) = 6.852 \approx 6.9$,
$\qquad s_{\overline{x}}^2 = 2^2(0.015) + ... + 11^2(0.029) - 6.852^2 = 3.546096 \approx 3.5461$, and
$\qquad s_{\overline{x}} \approx 1.88$.

$\qquad$ **(b)** $\overline{\overline{x}} \approx 6.9$ is close to $\mu = 7$ and $s_{\overline{x}} \approx 1.88$ is close to $\sigma/2 = \sqrt{14}/2 \approx 1.87$.

Section 6.1.3

6.13 Assuming a sample of size 25 is large enough to invoke the Central Limit Theorem, $P(\overline{X} > 20) \approx P[Z > 5(20 - 15)/10] = P(Z > 2.5) = 0.00621$.

6.15 **(a)** If the claim is true, the mean and standard deviation of the sample mean are 200 ohms and $2/6 = 1/3$ ohm, respectively. Since 198 ohms is 6 standard deviations below the mean, $P(\overline{X} \leq 198) \approx 0$.

$\qquad$ **(b)** Since obtaining a value that is 6 standard deviations below the mean is so unusual, we have reason to believe that the true average is less than the one stated.

6.17 If the process mean and standard deviation are 0.501 inch and 0.001 inch, respectively, $\overline{X}$ has mean and standard deviation 0.501 inch and $0.001/5 = 0.0002$, respectively. So, the probability that the test implies that the process is out of control is approximately

$$1 - P\left(0.4994 \le \overline{X} \le 0.5006\right) \approx 1 - P\left(\frac{0.4994-0.5010}{0.0002} \le Z \le \frac{0.5006-0.5010}{0.0002}\right)$$
$$= 1 - P(-8.00 \le Z \le -2.00)$$
$$\approx 1 - P(Z \le -2.00)$$
$$= 1 - 0.02275 = 0.97725.$$

6.19 (a) $E[\overline{X}] = E[X] = 0.50$ as before.

$Var(\overline{X}) = Var(X)/25 = 1/300$ is smaller than that in Problem 6.18.

(b) $P(\overline{X} > 0.67) \approx P[Z > \sqrt{300}(0.67 - 0.50)] \approx P(Z > 2.94) = 0.00164$. This is much smaller than the probability in Problem 6.18(a).

(c) Since $\overline{X}$ is approximately normal, the 10th percentile is approximately 1.28 standard deviations below the mean of the sampling distribution. So, $\overline{X}_{0.10}$ is approximately $0.50 - (1.28)(1/\sqrt{300}) \approx 0.43$. $\overline{X}_{0.10}$ is larger for $n = 25$ than for $n = 9$.

6.21 (a) $\overline{X}$ is approximately normal with mean 36 inches and standard deviation $0.60/\sqrt{100} = 0.06$ inch. Thus, $P(\overline{X} > 36.15) \approx$

$P[Z > (36.15 - 36.00)/0.06] = P(Z > 2.50) = 0.00621$.

(b) The standardized value of $36 + \Delta$ is approximately 1.645. So, $\Delta \approx (1.645)(0.06) \approx 0.099$.

(c) Since $P(Z > 1.96) \approx 0.025$, 36.15 is approximately 1.96 standard deviations above the mean. So, $\sqrt{n}(36.15 - 36.00)/0.60 \approx 1.96$. This implies that $n \approx [(1.96)(0.60)/0.15]^2 \approx 61.5$. Rounding up, a sample of size $n = 62$ is required.

Sections 6.2.1 and 6.2.2

6.23 $1 - \alpha = P\left[\mu - Z_p\left(\sigma/\sqrt{n}\right) \le \overline{X} \le \mu + Z_p\left(\sigma/\sqrt{n}\right)\right]$

$\qquad = P\left[\left(-\overline{X}\right) - Z_p\left(\sigma/\sqrt{n}\right) \le -\mu \le \left(-\overline{X}\right) + Z_p\left(\sigma/\sqrt{n}\right)\right]$

$\qquad = P\left[\overline{X} + Z_p\left(\sigma/\sqrt{n}\right) \ge \mu \ge \overline{X} - Z_p\left(\sigma/\sqrt{n}\right)\right]$

$\qquad = P\left[\overline{X} - Z_p\left(\sigma/\sqrt{n}\right) \le \mu \le \overline{X} + Z_p\left(\sigma/\sqrt{n}\right)\right]$

6.25 Let X denote the length of life of a randomly selected transistor. If $X \sim N(\mu, 900)$, $225 \pm 2.576\left(30/\sqrt{30}\right) \approx 225 \pm 14.1 = [210.9, 239.1]$ is a 99% confidence interval for μ.

6.27 Let X denote the manufacturing time. Since a sample of 49 typewriters is used, we invoke the Central Limit Theorem and assume that $\overline{X}$ is approximately normal. Also assuming that $\sigma \approx s = 6$, the interval $48.0 \pm 2.576\left[6/\sqrt{49}\right] \approx 48.0 \pm 2.2 = [45.8, 50.2]$ is an approximate 99% confidence interval for μ.

6.29 Since $\overline{x} \approx 0.0633$, $s \approx 0.0070$, and $n = 40$, $0.0633 \pm 1.96\left(0.0070/\sqrt{40}\right) \approx 0.0633 \pm 0.0022 = [0.0611, 0.0655]$ is an approximate 95% confidence interval for the average warpage.

(*Note*: A box plot reveals that 0.080 and 0.082 are outliers. A histogram indicates that the sampled distribution may be skewed to the right. So, the outliers may really represent the sampled population. Unless further investigation indicates some special cause was acting, the interval obtained here should be used with caution.)

Section 6.2.3

6.31 The correct choice is (b).

6.33 (a) $T_{0.995} = 2.797$ and $53.880 \pm 2.797\left(1.8102/\sqrt{25}\right) \approx 53.880 \pm 1.0126 \approx$ [52.9, 54.9]. We are 99% confident that the average temperature is at least 52.9 degrees and at most 54.9 degrees.

(b) $T_{0.95} = 1.711$ and $53.880 + 1.711\left(1.8102/\sqrt{25}\right) \approx 53.880 + 0.6195 \approx 54.5$. We are 95% confident that the average temperature does not exceed 54.5 degrees.

(c) $T_{0.90} = 1.318$ and $53.880 - 1.318\left(1.8102/\sqrt{25}\right) \approx 53.880 - 0.4772 \approx 53.4$. We are 90% confident that the average temperature is at least 53.4 degrees.

6.35 (a) Using $T_{0.95} = 1.833$, $\overline{x} = 17.1$, and $s = 3$, $17.1 + 1.833\left(3/\sqrt{10}\right) \approx 18.8$ is an upper 95% confidence limit. The manufacturer is 95% confident that the average emission does not exceed 18.8 ppm. Thus, we have strong evidence that the true population mean is less than 20 ppm.

(b) The assumptions are that the data were randomly obtained from a population that is normally distributed. (*Note*: Unless the process is stable over time, an interval such as that in (a) applies only to the population at the time of sampling.)

6.37 From Problem 6.24, [365.1, 384.9] is a 90% confidence interval for μ. Thus, 384.9 is a 95% upper confidence limit. This means that we are 95% confident that the average charge-to-tap time is at most 384.9 minutes.
(*Note*: $\overline{x} + Z_{0.95}\left(\sigma/\sqrt{n}\right) = 375 + (1.645)\left(29/\sqrt{23}\right) \approx 384.9$)

6.39 $\bar{x} \approx 29.233$, $s \approx 1.6121$, and exploratory analysis indicates that a normality assumption is reasonable. Confidence intervals and limits will be obtained using quantiles of a t distribution with 29 degrees of freedom.

   (a) Since $T_{0.975} = 2.045$, $29.233 \pm (2.045)(1.6121/\sqrt{30}) \approx 29.233 \pm 0.6019 \approx$ [28.6, 29.8] is a 95% confidence interval for μ. We are 95% confident that the average resistance is at least 28.6 Ω and at most 29.8 Ω.

   (b) Since $T_{0.99} = 2.462$, $29.233 + (2.462)(1.6121/\sqrt{30}) \approx 30.0$ is a 99% upper confidence limit for μ. We are 99% confident that the average resistance is at most 30.0 ohms.

   (c) Since $T_{0.98} = 2.150$, $29.233 - (2.150)(1.6121/\sqrt{30}) \approx 28.6$ is a 98% lower confidence limit for μ. We are 98% confident that the average resistance is at least 28.6 ohms.

Section 6.2.4

6.41 Using Equation (6.12) with $\Delta = 2$, $Z_{0.95} = 1.645$, and $\sigma = 9$, $n = [1.645(9)/2]^2 \approx 54.8$. Rounding up, a sample of size $n = 55$ is required.

6.43 Using Equation (6.12) with $\Delta = 0.8$, $Z_{0.995} = 2.576$, and $\sigma \approx s = 1.6121$, $n = [2.576(1.6121)/0.8]^2 \approx 26.9$. Rounding up, a sample of size 27 is required when σ is known to be 1.6121.

In this case, σ is unknown. For a 90% probability that the interval width will not exceed 1.6, we use Appendix M with $p^* = 0.90$ and $1 - \alpha = 0.99$. Since $n^* = 37$ when $n = 25$ and $n^* = 43$ when $n = 30$, we use linear interpolation and find $n^* \approx 40$ when $n = 27$. Thus, a sample of at least 40 items should be used.

Section 6.3

6.45 $\overline{X} - \overline{Y}$ is normally distributed with mean $15 - 12 = 3$ and variance $(1.69/9) + (2.25/16) \approx 0.3284$. Thus, $P(\overline{X} \geq \overline{Y} + 4) = P(\overline{X} - \overline{Y} \geq 4) = P[Z \geq (4-3)/\sqrt{0.3284}] \approx P(Z \geq 1.75) = 0.04006$.

6.47 Using Definition 6.4 with $Z_{0.99} = 2.33$, $(68.4 - 59.7) \pm 2.33\sqrt{(34.2/21) + (28.9/26)}$ $\approx 8.7 \pm 3.9 = [4.8, 12.6]$ is a 98% confidence interval for $\mu_X - \mu_Y$.

6.49 Suppose $\mu_A = \mu_B$. Since $s^2_{pooled} = [(6-1)(0.035) + (8-1)(0.200)]/(14-2) = 0.13125$, the observed value of T is $t \approx [(3.57 - 4.00) - 0]/\sqrt{0.13125(\frac{1}{6} + \frac{1}{8})} \approx$ -2.198.

6.51 If $m = n$, $S^2_{pooled} = \dfrac{(m-1)S^2_X + (m-1)S^2_Y}{m+m-2} = \dfrac{(m-1)\left(S^2_X + S^2_Y\right)}{2(m-1)} = \dfrac{S^2_X + S^2_Y}{2}$. So,

$$T = \frac{\left(\overline{X} - \overline{Y}\right) - \left(\mu_X - \mu_Y\right)}{\sqrt{\left(\dfrac{S^2_X + S^2_Y}{2}\right) \times \left(\dfrac{1}{m} + \dfrac{1}{m}\right)}} = \frac{\left(\overline{X} - \overline{Y}\right) - \left(\mu_X - \mu_Y\right)}{\sqrt{\dfrac{S^2_X}{m} + \dfrac{S^2_Y}{m}}} = T'.$$

6.53 **(a)** We are assuming that independent, random samples were obtained from normally distributed populations with equal variances. Comparison of the sample variances does not suggest equal population variances.

   Using Equation (6.23) with $T_{0.99} = 2.382$ the 99th percentile of a t distribution with 69 degrees of freedom,

$$(49.260 - 48.940) - 2.382\sqrt{\tfrac{(2.987)^2}{70} + \tfrac{(1.832)^2}{70}} \approx 0.320 - 0.998 = -0.678$$

is a 99% lower confidence limit for $\mu_2 - \mu_1$. With 99% confidence, $\mu_2 - \mu_1 \geq -0.678$.

(b) Since the 99% lower confidence limit is negative, $\mu_2 - \mu_1 = 0$ (or $\mu_2 < \mu_1$) is possible. The data do not indicate that the new radio has, on average, better quieting. This gives us no strong reason to change to the new type of radio.

6.55

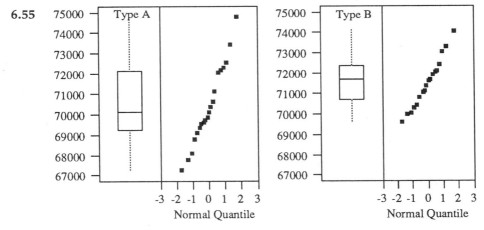

Consider the preceding JMP normal quantile plots and box plots. The quantile plots are reasonably linear and the box plots reveal no mild or strong outliers. The box plots and sample standard deviations seem to indicate that the population variances differ. The 95% confidence interval obtained using the Hsu solution, [-2,146.04, -64.44], provides a more reliable estimate than the interval obtained in Problem 6.54.

6.57 Consider Definition 6.4. We must determine the smallest value of n for which $Z_p \sqrt{\left(\sigma_X^2/n\right)+\left(\sigma_Y^2/n\right)} \leq \Delta$. For the situation presented here, $Z_p = Z_{0.975} = 1.96$, $\sigma_X = 12$, $\sigma_Y = 15$, and $\Delta = 3$. If $1.96\sqrt{\left(12^2+15^2\right)/n} \leq 3$, $\sqrt{n} \geq 1.96\sqrt{369}/3$. Squaring both sides, $n \geq (1.96^2)(369)/9 \approx 157.5$. Thus, a sample of 158 observations is required.

6.59 Using $R \approx 4\sigma$, $\sigma \approx 6/4 = 1.5$ for each model. Using Equation (6.21) with $Z_{0.975} = 1.96$, $\sigma = 1.5$, and $\Delta = 0.50$, $n = 2[(1.96)(1.5)/0.50]^2 \approx 69.1$. Rounding up, a common sample size of $m = n = 70$ should be used when σ is known to be 1.5.

In this case, σ is unknown. Using Appendix N with $1 - \alpha = 0.95$, $p^* = 0.90$, and $n = 70$, we find $n^* = 82$. Thus, random samples of at least 82 observations should be obtained from each population.

Section 6.4

6.61 The sample statistics $\bar{x} = 20$, $s^2 = SSX/8 = 288/8 = 36$, and $s = 6$ were observed.

    (a) Letting $n = 9$, $1 - \alpha = 0.95$, and $p = 0.95$, we find $K = 3.532$ in Table J-1 of Appendix J. Thus, $20 \pm (3.532)(6) \approx 20 \pm 21.2 = [-1.2, 41.2]$ is a 95% tolerance interval for 95% of the values of X. We are 95% confident that at least 95% of the values in the sampled population are between -1.2 and 41.2.

    (b) Letting $n = 9$, $1 - \alpha = 0.95$, and $p = 0.99$, we find $K = 4.143$ in Table J-2 of Appendix J. Thus, $20 + (4.143)(6) \approx 44.9$ is an upper 99% tolerance limit with 95% confidence. We are 95% confident that at least 99% of the values in the sampled population do not exceed 44.9.

6.63 **(a)** Using Table K-1 of Appendix K with $n = 24$ and $p = 0.90$, we find $1 - \alpha = 0.70$. The confidence level is 70%.

    (b) Using Table K-1 of Appendix K with $n = 24$ and $1 - \alpha = 0.95$, we find $0.80 < p < 0.85$. Linear interpolation gives $p \approx 0.81$.

6.65 The normal quantile plot in Figure 5.5 supports an assumption of normality. Using Table J-1 in Appendix J with $1 - \alpha = 0.90$, $p = 0.95$, and $n = 12$, we find $K = 2.863$. Since $\bar{x} = 120$ and $s \approx 21.3882$, $\bar{x} \pm Ks = 120 \pm (2.863)(21.3882) \approx 120 \pm 61.2 = [58.8, 181.2]$. We are 90% confident that at least 95% of the sampled values fall between 58.8 and 181.2.

6.67 **(a)** Consider the following JMP graphics. There are 3 outliers. These are associated with units 8, 21, and 27, which malfunctioned. They will be removed, and further analysis will involve only the remaining 29 data values.

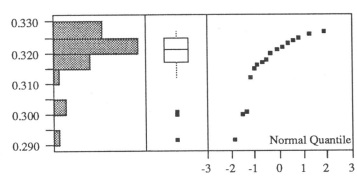

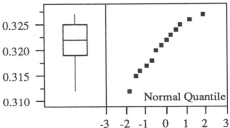

(b) A box plot and normal quantile plot of the edited data set are shown. The normal quantile plot is reasonably linear and the box plot reveals no outliers. A normality assumption seems reasonable.

(c) Assuming normality and using Table J-2 of Appendix J with $1 - \alpha = 0.99$ and $p = 0.999$, we find $K \approx 4.508$. Since the sample mean and standard deviation are 0.3215 and 0.0038, respectively, we are 99% confident that 99.9% of the security units have a quiescent current of at most $0.3215 + (4.508)(0.0038) \approx 0.3386$ milliamperes.

(*Note*: If you believe the evidence favors nonnormality, a distribution-free tolerance limit should be used. From Table K-2 in Appendix K, we find that $p = 0.85$ when $1 - \alpha = 0.99$ and $n = 29$.

The table does not include sample sizes required for a distribution-free, 99% upper tolerance limit for 99.9% of the units. Since $n = 919$ when $p = 0.995$ and $1 - \alpha = 0.99$, we can say that more than 919 units would be required.)

Section 6.5

6.69 X has a binomial distribution with $n = 400$ and $\theta = 0.02$.

(a) $E[X/400] = \theta = 0.02$

(b) $Var(X/n) = \theta(1 - \theta)/n = (0.02)(0.98)/400 = 0.000049 = (0.007)^2$

(c) $X/400 \approx N(0.02, 0.000049)$

(d) $P\left(\frac{X}{400} < 0.02014\right) \approx P\left(Z < \frac{0.02014 - 0.02000}{0.007}\right) \approx P(Z < 0.02) = 0.50798$

6.71 Since $715/1,100 = 0.650$, $0.650 \pm 1.96\sqrt{(0.650)(0.350)/1,100} \approx 0.650 \pm 0.028$ $= [0.622, 0.678]$ is an approximate 95% confidence interval for the true proportion of VCR owners who fast forward recorded programs to avoid

63

viewing commercials. If you were paying for a commercial that was shown during a program that was recorded by most of the viewers, it would be disconcerting to know that fewer than 40% of the potential audience would see your commercial.

6.73 (a) Since $\left(\frac{119}{174}\right) \pm 1.645\sqrt{\frac{\left(\frac{119}{174}\right)\left(\frac{55}{174}\right)}{174}} \approx 0.684 \pm 0.058 = [0.626, 0.742]$, we are 90% confident that at least 62.6%, but no more than 74.2%, of the wire bonds are good.

  (b) For $X \sim b(174, 0.742)$, $P(X \le 119) \approx 0.050$, and for $X \sim b(174, 0.621)$, $P(X \ge 119) \approx 0.050$. Therefore, $[0.621, 0.742]$ is a 90% confidence interval for θ.

6.75 Since $152/800 = 0.190$ and $0.190 \pm 1.645\sqrt{(0.190)(0.810)/800} \approx 0.190 \pm 0.023$, $[0.167, 0.213]$ is a 90% confidence interval for the true proportion. We are 90% confident that at least 16.7%, but no more than 21.3%, of all U.S. companies provide health club memberships for executive officers.

6.77 Using Equation (6.33) with $\theta = 0.50$, $Z_{0.975} = 1.96$, and $\Delta = 0.03$, $n = (0.50)(0.50)(1.96/0.03)^2 \approx 1,067.1$. Rounding up, a sample of 1,068 people should be used.

6.79 Using Equation (6.33) with $\theta = 0.50$, $Z_{0.975} = 1.96$, and $\Delta = 0.05$, $n = (0.50)(0.50)(1.96/0.05)^2 = 384.16$. Rounding up, a maximum of 385 observations are required.

  The process must be stable over time. Otherwise, the proportion of nonconforming parts obtained at one time will be of no value when estimating the proportion for a later (or earlier) time.

Section 6.6

6.81 (a) Using Definition 6.8 with $x = 24$, $m = 300$, $x/m = 0.08$, $y = 12$, $n = 200$, $y/n = 0.06$, and $Z_{0.975} = 1.96$, $(0.08 - 0.06) \pm 1.96\sqrt{\frac{(0.08)(0.92)}{300} + \frac{(0.06)(0.94)}{200}}$ $\approx 0.02 \pm 0.045$ is an approximate 95% confidence interval for $\theta_1 - \theta_2$.

  (b) For both processes, the proportions of broken candy canes must be stable over time. The trials must be independent. The normal approximations must be valid.

  (c) Since the confidence interval contains 0, the evidence does not support the engineer's claim.

6.83 Let θ_1 denote the proportion of allergy sufferers who experience drowsiness when taking Seldane-D and θ_2 denote that for those when taking Claritin. Assuming the samples were independently and randomly obtained from the universe of allergy sufferers,

$(0.080 - 0.072) \pm 1.96\sqrt{\frac{(0.080)(0.920)}{1926} + \frac{(0.072)(0.928)}{374}} \approx 0.008 \pm 0.029$

is an approximate 95% confidence interval for $\theta_2 - \theta_1$. Since this

64

interval contains 0, we have insufficient evidence to claim that fewer persons experience drowsiness when using one medication than when using the other.

6.85 Let $\theta_1 = \theta_2 = 0.10$, $\Delta = 0.05$, and $Z_{0.95} = 1.645$. Using the formula in Problem 6.84, $n = (1.645/0.05)^2[(0.10)(0.90) + (0.10)(0.90)] \approx 194.8$. Thus, at least 195 homes should be inspected for each contractor.

6.87 Let $\theta_1 = 24/300 = 0.08$, $\theta_2 = 12/200 = 0.06$, $\Delta = 0.01$, and $Z_{0.975} = 1.96$. Using the formula in Problem 6.84, $n = (1.96/0.01)^2[(0.08)(0.92) + (0.06)(0.94)] \approx 4{,}994.08$. Rounding up, at least 4,995 candy canes from each process should be inspected.

Section 6.7

6.89 **(a)** If $X \sim N(\mu, 6^2)$ and $W \sim \chi^2_{(19)}$, $P(S^2 > 100) = P\left(\frac{19S^2}{36} > \frac{19(100)}{36}\right) \approx$ $P(W > 52.8) \approx 0.00005$. (*Note*: Using Appendix F, $P(W > 38.582) = 0.005$. So, $P(W > 52.8) < 0.005$.]

(b) Let $W = 19S^2/36$. Since W has a chi-square distribution with 19 degrees of freedom, $P(W > 30.144) = 0.05$. Thus, $19s^2/36 = 30.144$ and $s^2 = 36(30.144)/19 \approx 57.115$.

(c) The sampled population must have a normal distribution.

6.91 **(a)** If W denotes a chi-square random variable with 29 degrees of freedom, $W_{0.005} = 13.121$ and $W_{0.995} = 52.336$. Using Definition 6.9 with these quantiles and $s^2 \approx 2.5989$, $[L, U] = \left[\frac{29(2.5989)}{52.336}, \frac{29(2.5989)}{13.121}\right] \approx$ $[1.440, 5.744]$ is a 99% confidence interval for σ^2. Thus, we are 99% confident that the population variance is at least 1.440 ohms^2 and at most 5.744 ohms^2.

(b)

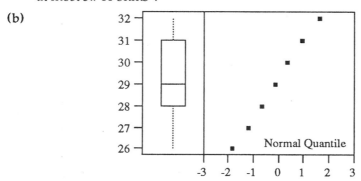

The preceding JMP box plot reveals no outliers or suspect outliers. The normal quantile plot is reasonably linear. Based on these results, an assumption of normality seems reasonable. The confidence interval obtained in (a) should be adequate.

6.93 **(a)** If W denotes a chi-square random variable with 8 degrees of freedom, $W_{0.025} = 2.180$ and $W_{0.975} = 17.535$. Using Definition 6.9 with these quantiles and $s^2 \approx 5{,}415{,}828$, $\left[\frac{8(5{,}415{,}828)}{17.535}, \frac{8(5{,}415{,}828)}{2.180} \right] \approx$ [2,470,865, 19,874,598] is a 95% confidence interval for σ^2, if the sample was randomly selected from a normally distributed population. Taking square roots of the interval endpoints for σ^2, a 95% confidence interval for σ is [1,571.9, 4,458.1].

(b) Since the lower endpoint of the confidence interval for σ is greater than 1,000, we have reason to believe that the standard deviation is greater than 1,000 pounds per square inch.

(c) A box plot and normal quantile plot are given to the right. For such a small sample size, there is no strong evidence favoring nonnormality. Based on these graphics, an assumption of normality seems reasonable. The results in (a) and (b) should be reliable.

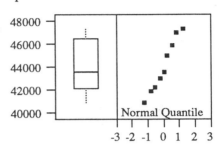

Section 6.8

6.95 **(a)** $P(F > 7.79) = 1 - P(F \leq 7.79) = 1 - 0.99 = 0.01$

(b) $P(F > 4.03) = 1 - P(F \leq 4.03) \approx 1 - 0.95 = 0.05$

(c) $P(F > 7.79) \approx 0.0019$ (*Note:* Using Appendix I, $P(F > 6.102) = 0.005$. Thus, $P(F > 7.79) < 0.005$.)

(d) $P(0.221 < F \leq 2.62) = P(F \leq 2.62) - P(F \leq 0.221) \approx 0.95 - 0.05 = 0.90$

6.97 **(a)** Assuming normality and independence, and using Definition 6.10 with $s_A^2 = 3{,}754{,}221$, $s_B^2 = 1{,}473{,}659$, $v_1 = v_2 = 20$, $F_{0.025} = 0.406$, and $F_{0.975} = 2.464$, $\left[\frac{3{,}754{,}221}{1{,}473{,}659} \times 0.406, \frac{3{,}754{,}221}{1{,}473{,}659} \times 2.464 \right] \approx [1.03, 6.28]$ is a 95% confidence interval for the ratio of the two population variances. We are 95% confident that $1.03\sigma_B^2 \leq \sigma_A^2 \leq 6.28\sigma_B^2$.

(b) Since the lower 95% confidence limit is greater than 1, we have reason to believe that the variance in the Type A steel is greater than that of the Type B steel.

(c) Box plots and normal quantile plots follow. There are no outliers, and the normal quantile plots are quite linear. Assuming normality seems reasonable. Therefore, the confidence interval in (a) should be reliable.

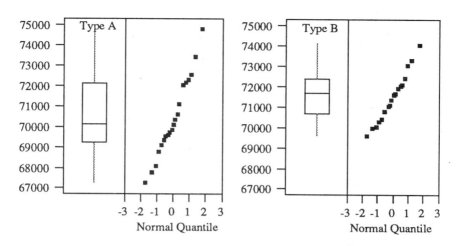

6.99 **(a)** Assuming normality and using Definition 6.10 with $s_A^2 = 380.7828$, $s_B^2 = 147.6103$, $v_1 = 29$, $v_2 = 29$, $F_{0.05} = 0.537$, and $F_{0.95} = 1.861$, $\left[\frac{380.7828}{147.6103} \times 0.537, \frac{380.7828}{147.6103} \times 1.861 \right] \approx [1.4, 4.8]$ is a 90% confidence interval for the ratio of the two population variances. We are 90% confident that $1.4\sigma_B^2 \leq \sigma_A^2 \leq 4.8\sigma_B^2$.

(b) No. Since the lower endpoint of the interval in (a) is greater than 1, the result in (a) indicates that the variability in x-chromaticity is greater for Vendor A than for Vendor B.

(c) JMP graphics for the two sets of data follow. The graphics for Vendor B reveal that 138 is an outlier. Since the outlier is not consistent with the other data, we will remove it and repeat the analysis. Notice, however, that the removal of that data value will reduce the variability and make the evidence for unequal variances even more compelling.

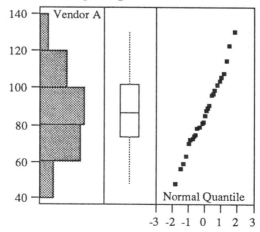

67

(*Note*: The exploratory analysis that identified the outlier should occur before the formal procedure. It should also occur while all parts are still available. This makes it more likely that a cause for the outlier will be identified.)

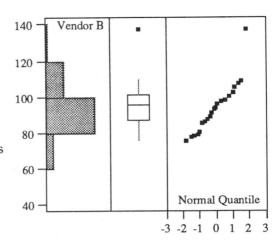

Since the outlier is so different from the other data, we will remove it and proceed with further analyses. Graphics of the modified data follow. Notice that the normal quantile plots are quite linear, with the slope greater for the Vendor A data. This implies that the x-chromaticity is more variable for Vendor A than for Vendor B. The box plots and histograms support that claim.

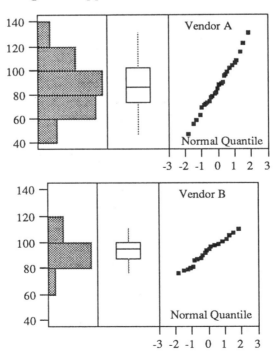

Since a normality assumption seems justified, we will estimate the ratio of the two population variances. The sample variance for the Vendor B modified data is 84.4138. Using Definition 6.10 with $s_A^2 = 380.7828$, $s_B^2 = 84.4138$, $v_1 = 28$, $v_2 = 29$, $F_{0.05} = 0.533$, and $F_{0.95} = 1.868$, $\left[\frac{380.7828}{84.4138} \times 0.533, \frac{380.7828}{84.4138} \times 1.868 \right] \approx [2.4, 8.4]$ is a 90% confidence interval for the ratio of the two population variances. We are 90% confident that $2.4\sigma_B^2 \leq \sigma_A^2 \leq 8.4\sigma_B^2$. In this case, our observations have created a stronger case for concluding that the x-chromaticity is more variable for Vendor A than for Vendor B.

An interval estimate for the difference between the two population means can be obtained using the Hsu solution. For 90% confidence, we use Equation (6.23) with $T_{0.95} = 1.701$ the 95th percentile of a t distribution with 28 degrees of freedom. The resulting 90% confidence interval is $(87.1 - 93.4) \pm 1.701\sqrt{\frac{19.5^2}{30} + \frac{9.2^2}{29}} \approx -6.3 \pm 6.7 = [-13.0, 0.4]$. Since this interval contains 0, we have insufficient evidence to conclude that the two sample means differ.

Section 6.9

6.101 (a) A table of sample ranges follows. The average range is $\overline{R} = \Sigma R_i / 18 = 4.06/18 \approx 0.226$. Since $d_2 = 2.326$ for samples of size 5, $\overline{R}/d_2 \approx 0.226/2.326 \approx 0.097$ is an unbiased estimate of σ_X.

Sample	R	Sample	R	Sample	R
1	0.15	7	0.11	13	0.16
2	0.32	8	0.47	14	0.16
3	0.30	9	0.23	15	0.20
4	0.18	10	0.25	16	0.29
5	0.36	11	0.19	17	0.16
6	0.18	12	0.24	18	0.11

(b) Since $\sigma_{\overline{X}} = \sigma_X/\sqrt{5}$, $(\overline{R}/d_2)/\sqrt{5} \approx 0.097/\sqrt{5} \approx 0.043$ is an unbiased estimate of $\sigma_{\overline{X}}$.

6.103

Sample	s	Sample	s	Sample	s
1	0.076	7	0.042	13	0.073
2	0.132	8	0.190	14	0.067
3	0.123	9	0.097	15	0.081
4	0.073	10	0.107	16	0.116
5	0.152	11	0.073	17	0.076
6	0.066	12	0.096	18	0.047

Using the preceding table of sample standard deviations, $\bar{s} = \Sigma s_i / 18 = 1.687/18 \approx 0.094$. Since $c_4 = 0.940$ for samples of size 5, $\bar{s}/c_4 \approx 0.094/0.940$

≈ 0.100 is an unbiased estimate of σ_X. Since $\sigma_{\overline{X}} = \sigma_X / \sqrt{5}$, $\left(\overline{s}/c_4 \right)/\sqrt{5} \approx$
$0.100/\sqrt{5} \approx 0.045$ is an unbiased estimate of $\sigma_{\overline{X}}$.

Chapter 6 Supplementary Problems

6.105 T has a t distribution with 35 degrees of freedom. Using Appendix H, $P(T < 1.052) = 0.85$.

6.107 $E[X] = \int_0^3 x f(x) dx = \int_0^3 \frac{2}{9} x^2 dx = \frac{2}{27} x^3 \Big|_0^3 = 2$

$Var(X) = E\left[X^2\right] - 2^2 = \int_0^3 x^2 \left(\frac{2}{9} x\right) dx - 4 = \int_0^3 \frac{2}{9} x^3 dx - 4 = \frac{1}{18} x^4 \Big|_0^3 - 4 = 0.5$

$E\left[\overline{X}\right] = E[X] = 2$, $Var\left(\overline{X}\right) = Var(X)/36 = 1/72$, and $\sigma_{\overline{X}} = 1/6\sqrt{2}$

By the Central Limit Theorem, $\overline{X}$ has an approximate normal distribution. Therefore,
$P\left(1.780 < \overline{X} < 2.292\right) \approx P\left[6\sqrt{2}(1.780 - 2) < Z < 6\sqrt{2}(2.292 - 2)\right] \approx$
$P(-1.87 < Z < 2.48) = \Phi(2.48) - \Phi(-1.87) = 0.99343 - 0.03074 =$
$0.96269 \approx 0.96$.

6.109 **(a)** Before a sample is selected, the probability of obtaining a sample for which the corresponding confidence interval contains the parameter is 0.95. So, the engineer can expect an average of 95% of the intervals to contain the true value.

(b) $Y \sim b(100, 0.95)$, so $P(92 \leq Y \leq 98)$ $= P(Y \leq 98) - P(Y \leq 91)$
$= 0.96292 - 0.06309 \approx 0.90$.

6.115 **(a)** The sample mean and standard deviation are (to the nearest 100th) 39.23 and 13.25, respectively. For a t distribution with 29 degrees of freedom, the 95th percentile is 1.699. Using Definition 6.2,
$$39.23 \pm (1.699)\left(13.25/\sqrt{30}\right) \approx 39.23 \pm 4.11 = [35.12, 43.34]$$
is a 90% confidence interval for μ.

(b)

The preceding histogram and box plot are reasonably symmetric for the sample size used. There are no outliers, and the distribution seems to be centered near 40.

(c) The preceding normal quantile plot is fairly linear. The 3 points at the top of the plot may indicate a distribution that is moderately skewed to the right.

(d) Since the graphics may be indicating that the population has a moderately skewed distribution, and the chi-square procedure is adversely affected by nonnormality, an interval estimate of the population variance may be unreliable.

CHAPTER 7
STATISTICAL TESTS OF HYPOTHESES

Sections 7.1.1 and 7.1.2

7.1 (a) $H_0: \mu = 410$ (b) Test Statistic: $Z = \sqrt{23}(\overline{X} - 410)/29$
         $H_a: \mu < 410$ Decision Rule: Reject H_0 if $z < -1.645$.

(c) Since $z = \sqrt{23}(375 - 410)/29 \approx -5.788$, reject H_0 and conclude that the processing change reduced the average charge-to-tap time.

7.3 (a) $H_0: \mu = 1.00$ (b) Test Statistic: $Z = \sqrt{50}(\overline{X} - 1.00)/S$
         $H_a: \mu \neq 1.00$ Decision Rule: Reject H_0 if $|z| > 2.576$.

(c) Since $z = \sqrt{50}(1.02 - 1.00)/0.04 \approx 3.536$, reject H_0. The data indicate that, at the time of sampling, the average length was not 1.00 inch.

(d) The test is based on the assumption that (i) a random sample was obtained from a normal population or (ii) the sample size is so large that normality is not a concern due to the Central Limit Theorem.

7.5 (a) $H_0: \mu = 880$; $H_a: \mu < 880$; Test Statistic: $Z = \sqrt{50}(\overline{X} - 880)/S$
      Decision Rule: Reject H_0 if $z < -2.326$.
      Analysis: Since $\overline{x} = 871$ and $s = 21$, $z \approx -3.030$. Therefore, reject H_0 and conclude that $\mu < 880$ pounds.

(b) Assumptions: For a sample size this large, independence is the assumption that must be satisfied. Also, the process must be stable over time and sampling must be from a single population.

7.7 $H_0: \mu = 23{,}500$; $H_a: \mu < 23{,}500$; Test Statistic: $Z = \sqrt{100}(\overline{X} - 23{,}500)/S$
      Decision Rule: Reject H_0 if $z < -1.96$ (i.e., $\alpha = 0.025$).
      Analysis: Since $\overline{x} = 23{,}000$ and $s = 3{,}900$, $z \approx -1.282$. There is insufficient evidence to reject H_0. (*Note*: The *p*-value is $P(Z < -1.282) \approx 0.10$. Thus, rejection is possible for any $\alpha \geq 0.10$.)

7.9 **(a)** $\bar{x} - 1.645\left(s/\sqrt{40}\right) = 43.6 - 1.645\left(1.3/\sqrt{40}\right) \approx 43.26$

(b) We are approximately 95% confident that $\mu > 43.26$. This indicates that, at the 5% significance level, $H_0 : \mu = 43$ can be rejected in favor of $H_a : \mu > 43$.

7.11 **(a)** $H_0 : \mu = 0.063$; $H_a : \mu < 0.063$

Test Statistic: $Z = \sqrt{40}\left(\overline{X} - 0.063\right)/S$

Decision Rule: Reject H_0 if $z < -2.326$.

Analysis: Since $\bar{x} \approx 0.0576$ and $s \approx 0.0032$, $z \approx -10.67$. This is overwhelming evidence that the new laminate has less average warpage than the old.

Note: The values 0.065 and 0.069 are outliers. These may be indicating that the sampled population is highly skewed. Since the sample size is so large, the inference obtained here should be reliable.

(b) p-value = $P(Z \le -10.67) \approx 0$. When $\mu = 0.063$, we will almost never obtain a value of Z at least as extreme as the one obtained using the given data.

Section 7.1.3

7.13 $H_0 : \mu = 10.5$; $H_a : \mu > 10.5$

Test Statistic: $T = \sqrt{8}\left(\overline{X} - 10.5\right)/S$; $\nu = 7$

Decision Rule: Reject H_0 if $t > 1.415$.

Analysis: Since $\bar{x} = 14.0$ and $s \approx 3.2$, $t \approx 3.09$. Thus, we conclude that the average time required to fill such orders exceeds 10.5 days. This result is valid if the sampled population is approximately normal and the sample was randomly selected.

Note: A graphical analysis indicates no outliers. A normal quantile plot is reasonably linear for such a small sample.

7.15 **(a)** $H_0 : \mu = 1.11$; $H_a : \mu > 1.11$

Test Statistic: $T = \sqrt{5}\left(\overline{X} - 1.11\right)/S$; $\nu = 4$

Decision Rule: Reject H_0 if $t > 3.747$.

Analysis: Since $\bar{x} = 1.13$ inches and $s = 0.02$ inch, $t \approx 2.236$. There is insufficient evidence to reject the null hypothesis that the average length is 1.11 inches.

(b) $\bar{x} - 3.747\left(s/\sqrt{5}\right) = 1.13 - 3.747\left(0.02/\sqrt{5}\right) \approx 1.096$.

(c) Since the 99% lower confidence limit is 1.096, we can say that $\mu \ge 1.096$, with 99% confidence. However, we cannot conclude that $\mu > 1.11$ inches. This result agrees with that in (a).

7.17 (a) $H_0: \mu = 20; H_a: \mu < 20$

Test Statistic: $T = \sqrt{10}\left(\overline{X} - 20\right)/S; \nu = 9$

Decision Rule: Reject H_0 if $t < -1.833$.

(b) Since $\overline{x} = 17.1$ ppm and $s = 3$ ppm, $t \approx -3.057$. We conclude that the average emission level is less than 20 ppm.

(c) In Problem 6.35(a), we concluded that the true population mean is less than 20 ppm. We reached the same conclusion here.

7.19 (a) Let x denote an observed value for the elongation. To use the MEANS procedure to obtain the test results, coded data of the form $y = x - 40$ were entered. Thus, MEAN denotes $\overline{y}$ and STD ERROR OF MEAN denotes $s/\sqrt{12}$.

(b) p-value = PRT > $|T|$ = 0.0306 is the observed significance level for $H_0: \mu_Y = 0$ versus $H_a: \mu_Y \neq 0$. This test is equivalent to the test in the problem statement. Thus, when the average elongation is 40% per 2 inches, the probability of observing a value of T as extreme (in either direction) as -2.48 is approximately 0.0306.

(c) Since the p-value exceeds 0.02, the null hypothesis should not be rejected. Rejection occurs when the p-value is less than or equal to α.

(d) $\overline{x} \pm t_{0.99}\left(s/\sqrt{12}\right) = \left(\overline{y} + 40\right) \pm 2.718(1.11) = \left(-2.75 + 40\right) \pm 2.718(1.11) \approx$ 37.25 $\pm$ 3.02. The 98% confidence interval, [34.23, 40.27], includes 40. As in (c), we have insufficient evidence to conclude (at the 2% significance level) that $\mu \neq 40$.

7.21 $\overline{x} \pm t_{0.995}\left(s/\sqrt{30}\right) = 0.8383 \pm 2.756\left(0.1859/\sqrt{30}\right) \approx 0.8383 \pm 0.0935 =$ [0.7448, 0.9318] is a 99% confidence interval for μ. Since 0.85 is included in that interval, we have insufficient evidence to conclude that $\mu \neq 0.85$. This agrees with the conclusion of 7.20(a).

7.23 (a) $H_0: \mu = 21; H_a: \mu < 21$

Test Statistic: $T = \sqrt{15}\left(\overline{X} - 21\right)/S; \nu = 14$

Since $\overline{x} = 20$ and $s = 0.60$, $t \approx -6.455$. The p-value is $P(T \leq -6.455) \approx$ 0.0000. When $\mu = 21$, it is very unlikely that T will attain a value at least as extreme as -6.455.

(b) Since the p-value is almost 0, we can reject the null hypothesis for any reasonable value of α. Rejection occurs whenever the p-value is less than or equal to α. In this case, $\alpha = 0.05$, so we conclude that $\mu < 21$.

(c) In Example 6.18, 20.30 was found to be an upper 95% confidence limit for μ. Since this limit is less than 21, we can conclude (at the 5% level) that $\mu < 21$.

7.25 The correct choice is (b). [*Note*: A Type I error occurs when a true null hypothesis is rejected.]

7.27 You cannot make a Type II error when the null hypothesis is true. Thus, the probability is 0.

7.29 When the *p*-value exceeds α, the test statistic is not in the critical region. Thus, we cannot reject the null hypothesis.

7.31 If $\alpha = 0.01$, we conclude that $\mu < 410$ when the observed sample mean is less than $410 - 2.326(29/\sqrt{23}) \approx 395.9$ minutes. Thus, $\beta(381) =$
$$P\left(\overline{X} > 395.9 \,|\, \mu = 381\right) = P\left(Z > \sqrt{23}\left(395.9 - 381\right)/29\right) \approx P(Z > 2.46) = 0.00695.$$

7.33 Let $d = 2\sigma/\sigma = 2$. For this two-tailed test, we use Figure L-1 of Appendix L with $n = 16$, $\alpha = 0.05$, and $d = 2$ to find $\beta(2) \approx 0$. Therefore, we will almost always detect a shift of this magnitude.

7.35 Let $d = 0.5\sigma/\sigma = 0.5$. To determine n for a two-tailed test with $\alpha = 0.05$ and $\beta = 0.10$, we use Figure L-1 of Appendix L. The vertical line through $d = 0.5$ and the horizontal line through $\beta(d) = 0.10$ intersect in a point near the curve for $n = 40$. Thus, a sample of about 40 values is needed.

7.37 Using Equation (7.13) with $\alpha = 0.10$, $\sigma = 30$, $\delta = 5$, $\beta = 0.05$, and $Z_{0.05} = -1.645$, we find $n = \left[(30/5)\left((-1.645) + (-1.645)\right)\right]^{2} \approx 390$.

7.39 Using Equation (7.12) with $\alpha = 0.02$, $\sigma \approx s = 0.30$, $\delta = 0.20$, $\beta = 0.10$, $Z_{0.02} = -2.054$, and $Z_{0.10} = -1.282$, we find $n = \left[(0.30/0.20)\left((-2.054) + (-1.282)\right)\right]^{2} \approx$ 26. Since σ is unknown, increase the sample size to $26 + 3 = 29$.

7.41 Using Equation (7.13) with $\alpha = 0.05$, $\delta = \sigma/2$, $\beta = 0.10$, $Z_{0.025} = -1.960$, and $Z_{0.10} = -1.282$, we find $n = \left[2\left((-1.960) + (-1.282)\right)\right]^{2} \approx 43$.

7.43 **(a)** Using Equation (7.12) with $\alpha = 0.05$, $\sigma = 0.225$, $\delta = 0.1$, $\beta = 0.15$, $Z_{0.05} = -1.645$, and $Z_{0.15} = -1.036$, $n = \left[(0.225/0.1)\left((-1.645) + (-1.036)\right)\right]^{2} \approx 37$.

(b) Since the test is at the 5% level and σ is estimated by 0.225, we increase the result in (a) by 2. A sample size of about 39 is needed.

7.45 Using Equation (7.13) with $\alpha = 0.05$, $\sigma = 0.04$, $\delta = 0.024$, $\beta = 0.50$, $Z_{0.025} = -1.960$, and $Z_{0.50} = 0.00$, we find $n = \left[(0.040/0.024)\left((-1.96) + 0\right)\right]^{2} \approx 11$.

Section 7.2

7.47

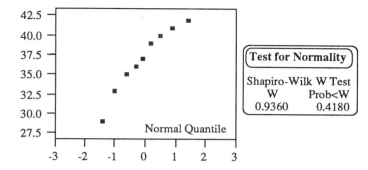

(a) The points in the preceding normal quantile plot form a reasonably linear pattern. The Shapiro-Wilk W Test has a moderately large p-value. Combining the two results, an assumption of normality seems reasonable.

(b) Since exploratory analysis indicates that a normality assumption is reasonable, the t procedure used in Problem 7.19 seems appropriate.

7.49

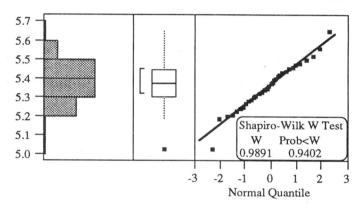

The histogram is reasonably bell-shaped. However, the box plot indicates that 5.03 is an outlier. When the histogram and normal quantile plot are considered, the value 5.03 seems to come from a long, lower tail of the distribution. Since the Shapiro-Wilk W test does not indicate that the hypothesis of normality should be rejected, the normal quantile plot is reasonably linear, and the histogram is reasonably symmetric and bell-shaped, a normality assumption seems plausible.

7.51 The following normal quantile plot is reasonably linear and the p-value of the Shapiro-Wilk W test is moderately large. The box plot indicates no outliers or suspect outliers. An assumption that the data were obtained from a normal distribution seems reasonable.

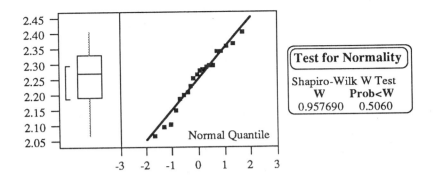

Test for Normality

Shapiro-Wilk W Test

W	Prob<W
0.957690	0.5060

Normal Quantile

Section 7.3.1

7.53 **(a)** Test Statistic: $T = (\overline{X}_1 - \overline{X}_2)/\sqrt{S^2_{pooled}[(1/10) + (1/10)]}$

$= (\overline{X}_1 - \overline{X}_2)/\sqrt{(S_1^2 + S_2^2)/10}$; $\nu = 18$

Decision Rule: Reject H_0 if p-value $\le \alpha$. We will use $\alpha = 0.05$.

Analysis: $t = (520 - 480)/\sqrt{(400 + 625)/10} \approx 3.951$

p-value $= P(T \ge 3.951) \approx 0.0005$.

Therefore, reject H_0 and conclude that $\mu_1 > \mu_2$.

(b) $(\overline{x}_1 - \overline{x}_2) - T_{0.995}\sqrt{(s_1^2 + s_2^2)/10} \approx (520 - 480) - 2.878\sqrt{(400 + 625)/10} \approx$

40 - 29.1 = 10.9. We are 99.5% confident that the average resistance of the Vendor 1 resistors exceeds that of the Vendor 2 resistors by at least 10.9 ohms.

7.55 **(a)** Test Statistic: $T = (\overline{X}_1 - \overline{X}_2)/\sqrt{S^2_{pooled}[(1/6) + (1/5)]}$; $\nu = 9$

Decision Rule: Since $T_{0.90} = 1.383$ for a t distribution with 9 degrees of freedom, reject H_0 if $t > 1.383$.

Analysis: $\overline{x}_1 = 0.141$, $s_1^2 \approx 0.0000056$, $\overline{x}_2 = 0.138$, and $s_2^2 \approx 0.0000085$

$s^2_{pooled} = [5(0.0000056) + 4(0.0000085)]/9 = 0.0000620/9$

$t = (0.141 - 0.138)/\sqrt{(0.0000620/9)[(1/6) + (1/5)]} \approx 1.888$

Therefore, reject H_0 and conclude that $\mu_1 > \mu_2$.

(b) p-value $= P(T \ge 1.888) \approx 0.046$. Since $0.046 < \alpha$, we can reject H_0. Further, only about 46 times in 1,000 will a value of $T \ge 1.888$ be obtained when the two means are equal.

Sections 7.3.2 and 7.3.3

7.57 Let $\sigma = 1.20$, $\delta = 1.80$, $\alpha = 0.05$, $Z_{\alpha/2} = Z_{0.025} = -1.96$, $\beta = 0.10$, and $Z_{\beta} = -1.282$. Using Equation (7.21) with $Z_{\alpha/2}$ substituted for Z_{α} gives

$$2\left[(1.20/1.80)((-1.96)+(-1.282))\right]^2 \approx 10 \text{ as the common sample size.}$$

(*Note*: If σ is unknown, but is estimated to be 1.20, we increase the sample size to 11.)

7.59 Using Equation (7.21) with $\sigma = 12$, $\delta = 1.50$, $\alpha = 0.05 = \beta$, and $Z_{\alpha} = Z_{\beta} = Z_{0.05} = -1.645$, we find $m = n = 2[(12/1.50)(-1.645)(2)]^2 \approx 1386$. Since σ is unknown and $\alpha = 0.05$, we increase this to 1387. Thus, two samples of about 1390 observations each should be used.

Section 7.3.4

7.61 In Problem 6.55, we decided that the two sampled populations appear to be normally distributed with unequal variances. Thus, we will use the Hsu solution.

$H_0: \mu_1 = \mu_2$; $H_a: \mu_1 \neq \mu_2$

Test Statistic: $T' = \left(\overline{X}_1 - \overline{X}_2\right)\Big/\sqrt{\left(S_1^2/21\right)+\left(S_2^2/21\right)}$; $\nu = 20$

Decision Rule: Reject H_0 if p-value $\leq \alpha$. (We will use $\alpha = 0.05$.)

Analysis: $\overline{x}_1 = 70{,}527.14$, $s_1 = 1937.581$, $\overline{x}_2 = 71{,}632.38$, $s_2 = 1213.944$, and

$$t' = (70{,}527.14 - 71{,}632.38)\Big/\sqrt{\left(1{,}937.581^2/21\right)+\left(1{,}213.944^2/21\right)} \approx$$

-2.215. Therefore, the p-value is $2P(T' \leq -2.215) \approx 0.04$. We reject $H_0: \mu_1 = \mu_2$ and conclude that the average tensile strength of I beams made from steel A is not equal to that of I beams made from steel B.

7.63 (a) For $\delta = 0.005$ and a common σ^2 of 0.00003,

$d = 0.005\Big/\left(\sqrt{0.00003}\sqrt{2}\right) \approx 0.65$. Using Figure L-4 in Appendix L, $\beta(0.65) \approx 0.07$. So, the probability of detecting a difference of 0.005 inch in the population means is approximately 0.93.

(b) Using Figure L-4 of Appendix L with $d = 0.65$ and $\beta(d) = 0.10$, a common sample size of about 35 is required. This sample size is valid if the two populations are normally distributed with equal variances. Exploratory analysis reveals that neither assumption seems reasonable.

7.65 From Problem 7.64, $s^2_{pooled} = 646.8$. We will assume the two populations have a common standard deviation of $\sigma \approx \sqrt{646.8} \approx 25.4$.

(a) Letting $\delta = 30$, $d = \delta/(\sigma\sqrt{2}) \approx 30/(25.4\sqrt{2}) \approx 0.8$. Using Figure L-3 with $n = 5$, $\beta(0.8) \approx 0.57$. Thus, the probability of concluding that $\mu_1 < \mu_2$ when $\mu_2 - \mu_1 = 30$ and $n = 5$ is approximately 0.43.

(b) Letting $\delta = 20$, $d = \delta/(\sigma\sqrt{2}) \approx 20/(25.4\sqrt{2}) \approx 0.6$. The point representing the intersection of the vertical line through 0.6 on the d axis and the horizontal line through 0.10 on the $\beta(d)$ axis is between the curves associated with $n = 20$ and $n = 30$. It appears that the point would be on the OC curve associated with $n = 27$. Thus, samples of about 30 fuses from each population are required. An additional 25 fuses should be selected from each.

Section 7.3.5

7.67 (a) Since before and after measurements were obtained on the same radio, the t procedure based on the differences in paired data should be used. Let X denote the reading for a radio before vibration, Y denote the reading for that same radio after vibration, and $D = Y - X$.

$H_0: \mu_D = 0$; $H_a: \mu_D > 0$

Test Statistic: $T = \sqrt{10}(\overline{D})/S_D$; $\nu = 9$

Decision Rule: Reject H_0 if the p-value does not exceed α. (We will use 0.05.)

Analysis: The sample of differences, {3, 4, 1, 2, -1, -1, 3, 4, 5, 5}, has mean and standard deviation (to nearest 100th) 2.50 and 2.22, respectively. Thus, $t \approx 3.56$ and p-value $= P(T \geq 3.56) \approx 0.0031$. Since $0.0031 < 0.05$, we conclude that $\mu_D > 0$. This is equivalent to concluding that $\mu_Y > \mu_X$. Hence, we have significant evidence of deterioration in the average performance following the vibration test.

(b)

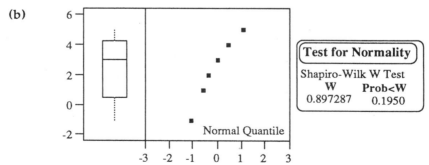

The normal quantile plot is reasonably linear. The p-value for a Shapiro-Wilk W Test is near 0.20. These two results indicate that an assumption of normality is reasonable.

78

(c) A scatter plot with least squares line follows. The scatter plot reveals a strong, linear relationship between the before and after readings.

The least squares line has equation $\hat{y} \approx 4.31 + 0.86x$. Since $r^2 \approx 0.76$, 76% of the variability in the readings after vibration is accounted for by this line.

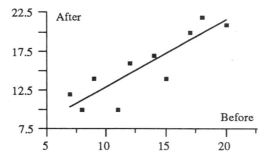

7.69 Let X and Y denote the yield strength of structures and bars, respectively.

(a) Since two strength readings are taken for each heat (one for structures, the other for bars), we will not assume the two samples are independent. Thus, we will analyze the differences between the paired observations. Let $D = Y - X$. The 10 observed values of D are 2,460; 1,100; -100; 9,850; 940; 5,140; -500; 60; 5,440; and -1,500.

$H_0: \mu_D = 0; H_a: \mu_D \neq 0$

Test Statistic: Assuming normality, $T = \sqrt{10}\left(\overline{D}\right)/S_D; \nu = 9$

Decision Rule: Using $\alpha = 0.05$, reject H_0 if $t < -2.262$ or $t > 2.262$.

Analysis: Since $\overline{d} = 2,289$ and $s_D \approx 3516.6$, $t \approx \sqrt{10}(2,289)/3,516.6 \approx 2.058$. This value is not in the critical region, so we have insufficient evidence to reject H_0. Thus, we do not reject the hypothesis that the average yield strengths of structures and bars are equal.

(b) Assuming normality, $2,289 \pm (2.262)(3,516.6)/\sqrt{10} \approx [-226.4, 4804.4]$ is a 95% confidence interval for μ_D. Since this interval contains 0, the two population means may be equal.

(*Note*: The following normal quantile plot of the sample of differences gives a strong impression that the sample data were obtained from a population that is skewed to the right. The *p*-value of the Shapiro-Wilk *W* Test is reasonably small. These facts seem to indicate that the sampled population may not be normally distributed. In such a case, the signed rank test discussed in Section 7.7.1 should be used. Results based on *t* procedures may be unreliable.)

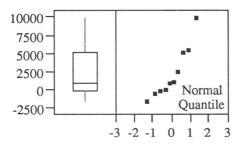

7.71 Since the values for a pair of observations obtained by a given person may be correlated, we will not assume independence. Let X denote the average of the 3 determinations obtained using the micrometer caliper and Y denote that of the vernier caliper. We will analyze $D = X - Y$, the differences in the paired observations.

(a) $H_0: \mu_D = 0; H_a: \mu_D \neq 0$

Test Statistic: $T = \sqrt{16}(\overline{D})/S_D = 4\overline{D}/S_D; \nu = 15$

Decision Rule: Using $\alpha = 0.025$, reject H_0 if p-value ≤ 0.025.
Analysis: The 16 observed values of D are 0, -0.002, 0, 0, 0.001, -0.002, 0.001, -0.006, -0.002, -0.003, 0, -0.003, -0.006, -0.007, 0.001, and -0.003. Since $\overline{d} = -0.0019$ and $s_D = 0.0026$ (to four decimal places), $t \approx 4(-0.0019)/0.0026 \approx -2.923$. The p-value is $2P(T \leq -2.923) \approx 0.01$. Thus, we have sufficient evidence to conclude that there is a difference between the means of the sampled populations.

(b) The following normal quantile plot seems to be slightly S-shaped, but reasonably linear. The p-value of the Shapiro-Wilk W Test is small. Combining these observations with the fact that the box plot is not symmetric, an assumption of normality seems questionable. Furthermore, the sample size is small. In such cases, use of the one-sample sign test (see Section 7.7.1) is recommended.

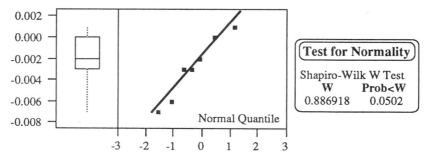

(c) Use of pairs of observations for each of the sixteen persons helps in two ways. First, it expands the inference space. If only one person had been used to obtain the measurements, we could only make claims about the difference in measurements obtained by that person. Second, it increases the sensitivity of the test by reducing the estimate in the variability due to the two measuring devices.

The average of two or more measurements is often used instead of a single measurement because the mean of a random sample selected from the distribution of a random variable varies less about the mean of that variable than a single, random value. However, the 3 values used to determine the sample mean provide information about the variability in the measurements that can be attributed to the persons making the measurements. That information is lost, if only the mean of the sample values is recorded. This is considered in detail in Chapter 9.

Sections 7.4.1 and 7.4.2

7.73 Since $x = 27$, $n = 14{,}256$, $n - x = 14{,}229$, and $Z_{0.02} \approx 2.054$,

$$\frac{27}{14{,}256} + 2.054\sqrt{\frac{(27/14{,}256)(14{,}229/14{,}256)}{14{,}256}} \approx 0.0026$$ is an approximate 98% upper confidence limit for the proportion of new radios that are nonconforming. Since this limit is less than 0.01, we conclude that the proportion of new radios that are nonconforming is less than that for the old radios.

7.75 Let X denote the number of nonconforming bulbs in a random sample of 600 bulbs taken from a week's production. Let θ denote the true proportion of nonconforming bulbs in that week's production. Assume $X \sim b(600, \theta)$.

$H_0: \theta = 0.03$; $H_a: \theta < 0.03$

Test Statistic: $Z = \dfrac{(X/600) - 0.03}{\sqrt{(0.03)(0.97)/600}} = \dfrac{X - 18}{\sqrt{17.46}}$

Decision Rule: Reject H_0 if p-value $< \alpha$. (We will use $\alpha = 0.05$.)
Analysis: Since $x = 12$, $z = (12 - 18)/\sqrt{17.46} \approx -1.436$. The p-value, $P(Z \le -1.436) = 0.07550$, is greater than α. There is insufficient evidence to reject H_0.

7.77 Since $x = 12$, $n = 600$, $n - x = 588$, and $Z_{0.95} \approx 1.645$, $\dfrac{12}{600} + 1.645\sqrt{\dfrac{(12/600)(588/600)}{600}} \approx$ 0.03 is an approximate 95% upper confidence limit for the proportion of nonconforming bulbs in the sampled week's production. We are approximately 95% confident that the week's production contains at most 3% nonconforming bulbs.

7.79 Let X denote the number of persons in a random sample of 1,000 employed Americans who work at least 2 jobs. Let θ denote the true proportion of employed Americans who work at least 2 jobs. Assume $X \sim b(1{,}000, \theta)$.
(a) $H_0: \theta = 0.05$; $H_a: \theta > 0.05$

Test Statistic: $Z = \dfrac{(X/1{,}000) - 0.05}{\sqrt{(0.05)(0.95)/1{,}000}} = \dfrac{X - 50}{\sqrt{47.5}}$

Decision Rule: Reject H_0 if p-value $< \alpha$. (We will use $\alpha = 0.05$.)

Analysis: Since $x = 62$, $z = (62 - 50)/\sqrt{47.5} \approx 1.74$. The p-value, $P(Z \geq 1.74) \approx 0.04$, is less than α. We conclude that, at the time of sampling, more than 5% of employed Americans worked 2 jobs.

(b) If $\alpha = 0.05$, the decision rule is "Reject H_0 if $(x - 50)/\sqrt{47.5} > 1.645$." This is equivalent to "Reject H_0 if $x > 50 + (1.645)\sqrt{47.5} \approx 61.34$." So,
$$\beta(0.054) = P[X \leq 61.34 \mid X \sim b(1{,}000, 0.054)]$$
$$= P[X \leq 61 \mid X \sim b(1{,}000, 0.054)] \approx 0.85.$$

(c) For $\alpha = \beta = 0.05$, $Z_\alpha = Z_\beta = -1.645$. Let $\theta_0 = 0.050$ and $\theta_1 = 0.054$. Using Equation (7.29) and rounding up,
$$n = \frac{\left[(-1.645)\sqrt{(0.050)(0.950)} + (-1.645)\sqrt{(0.054)(0.946)}\right]^2}{(0.050 - 0.054)^2} \approx 33{,}336.$$

Sections 7.4.3 and 7.4.4

7.81 $H_0: \theta = 0.50$; $H_a: \theta \neq 0.50$
Test Statistic: X, where $X \sim b(50, 0.50)$.
Decision Rule: Reject H_0 if p-value < 0.05.
Analysis: Since $x = 15$ is less than $n\theta_0 = 50(0.50) = 25$, the observed significance level is PROB $= 2P(X \leq 15) \approx 0.0066$. Therefore, we reject H_0 and conclude that $\theta \neq 0.50$.

7.83 Let X denote the number of nonconforming radios in a sample of 14,256 radios and θ denote the probability that a randomly selected radio is nonconforming. Binomial probabilities will be calculated using the **Binomial** routine in the **Statistical Tables** option in the **Calculator** module of StataQuest.
(a) $H_0: \theta = 0.01$; $H_a: \theta < 0.01$
Test Statistic: X, where $X \sim b(14{,}256, 0.01)$.
Decision Rule: Reject H_0 if p-value ≤ 0.02.
Analysis: Since $x = 27$, the observed significance level is PROB $= P(X \leq 27) \approx 0.0000$. As before, reject H_0 and conclude that the new radio has fewer nonconforming units than the old radio.
(b) Assumptions: $X \sim b(n, \theta)$ during the study and into the near future.
(c) When $X \sim b(14{,}256, 0.01)$, $P(X \leq 118) \approx 0.0191$ and $P(X \leq 119) \approx 0.0237$. Thus, to have $\alpha \leq 0.02$, the critical value of X is $c = 118$. So, $\beta(0.005) = 1 - P(X \leq 118 \mid X \sim b(14{,}256, 0.005)) \approx 1 - 1.0000 = 0.0000$, using the summary in Table 7.10 This result agrees with $\beta(0.005)$ obtained using a normal approximation.

7.85 Let X denote the number of nonconforming parts in a sample of 250 parts. Let θ denote the probability that a randomly selected part is nonconforming. Assume $X \sim b(250, \theta)$.
(a) $H_0: \theta = 0.04$; $H_a: \theta < 0.04$
Test Statistic: X, where $X \sim b(250, 0.04)$
Decision Rule: Reject H_0 if p-value < 0.05.

Analysis: Since $x = 3$, the observed significance level is PROB = $P(X \le 3) \approx 0.0093$. As before, reject H_0 and conclude that fewer than 4% of the parts in the lot are nonconforming.

(b) For $X \sim b(250, 0.03072)$, $P(X \le 3) \approx 0.0500$. Thus, 0.03072 is a 95% upper confidence limit for the proportion of nonconforming parts in the sampled lot.

(c) When $X \sim b(250, 0.04)$, $P(X \le 5) \approx 0.0633$) and $P(X \le 4) \approx 0.0270$. Thus, to have $\alpha \le 0.05$, the critical value of X is $c = 4$. Thus, $\beta(0.03) = 1 - P(X \le 4 \mid X \sim b(250, 0.03)) \approx 1 - 0.1282 = 0.8718$.

7.87 Let X denote the number of good bonds in a random sample of 174 wire bonds. Let θ denote the true proportion of good wire bonds in the sampled population.

   (a) $H_0: \theta = 0.65$; $H_a: \theta > 0.65$

      Test Statistic: X, where $X \sim b(174, 0.65)$.

      Decision Rule: Reject H_0 if p-value < α. (We will use $\alpha = 0.05$.)

      Analysis: Since $x = 119$, PROB = $P[X \ge 119 \mid X \sim b(174, 0.65)] \approx 0.1960$. As before, there is insufficient evidence to reject H_0.

   (b) When $X \sim b(174, 0.65)$, $P(X \ge 123) \approx 0.0661$ and $P(X \ge 124) \approx 0.0476$. Since 0.0476 is closer to 0.05, we choose $c = 124$ as the critical value of X. In this case, $P[X \ge 124 \mid X \sim b(174, 0.75)] \approx 0.8886$ is the probability of rejecting H_0 when 75% of the wire bonds are good.

7.89 Let X denote the number of kills in 300. Let θ denote the probability of a kill.

   (a) $H_0: \theta = 0.80$; $H_a: \theta > 0.80$

      Test Statistic: X, where $X \sim b(300, 0.80)$.

      Decision Rule: Reject H_0 if p-value < 0.04.

      Analysis: Since $x = 250$, the observed significance level is PROB = $P(X \ge 250) \approx 0.0830$. There is insufficient evidence to reject H_0.

   (b) When $X \sim b(300, 0.80)$, $P(X \ge 251) \approx 0.0622$, $P(X \ge 252) \approx 0.0457$, and $P(X \ge 253) \approx 0.0328$. Since 0.0457 is closest to 0.04, we choose $c = 252$ as the critical value of X. Thus, the probability of concluding that the probability of a kill exceeds 0.80 when $\theta = 0.85$ is $P[X \ge 252 \mid X \sim b(300, 0.85)] \approx 0.7186$.

   (c) The preceding results are based on the assumption that X is the sum of the outcomes in a sequence of 300 independent Bernoulli trials for which the probability of success is constant.

Section 7.5

7.91 Let X and Y denote the numbers of parts having open circuits in random samples of 150 parts from the lots of vendors 1 and 2, respectively.

$H_0: \theta_1 = \theta_2$; $H_a: \theta_1 > \theta_2$

Test Statistic: $Z = \left[\frac{X}{150} - \frac{Y}{150}\right] \Big/ \sqrt{\hat{\theta}(1 - \hat{\theta})\left[\frac{1}{150} + \frac{1}{150}\right]}$, with $\hat{\theta} = (X + Y)/300$

Decision Rule: Reject H_0 if p-value $\le \alpha$. (We will use $\alpha = 0.01$.)

Analysis: Since $x = 23$ and $y = 8$, $z = (15/150)\big/\sqrt{(31/300)(269/300)(2/150)} \approx$ 2.845 and the p-value is $P(Z \geq 2.845) \approx 0.00222$. Therefore, we conclude that the lot supplied by vendor 2 contains fewer parts with open circuits than does the lot supplied by vendor 1.

7.93 Let X and Y denote the numbers of good parts in random samples of 174 parts selected from those produced using the old and new processes, respectively. Let θ_1 and θ_2 denote the true proportions of good parts produced using the old and new processes, respectively. We assume that $X \sim b(174, \theta_1)$ and $Y \sim b(174, \theta_2)$ are appropriate statistical models and that normal approximations are adequate.

$H_0: \theta_1 = \theta_2 \; ; H_a: \theta_1 < \theta_2$

Test Statistic: $Z = \left[\frac{X}{174} - \frac{Y}{174}\right]\big/\sqrt{\hat{\theta}(1 - \hat{\theta})\left[\frac{1}{174} + \frac{1}{174}\right]}$, with $\hat{\theta} = (X + Y)/348$

Decision Rule: Reject H_0 if p-value ≤ 0.05.

Analysis: Since $x = 119$ and $y = 147$, $z = \left(\frac{-28}{174}\right)\big/\sqrt{\left(\frac{266}{348}\right)\left(\frac{82}{348}\right)\left(\frac{2}{174}\right)} \approx -3.54$ and the p-value is $P(Z \leq -3.54) = 0.00020$. We conclude that the proportion of good parts produced by the new process is greater than that produced by the old process.

7.95 Let $\alpha = 0.05$, $Z_\alpha = -1.645$, $\beta = 0.10$, $Z_\beta = -1.282$, $\theta_1 = 0.046$, $\theta_2 = 0.026$, and $\delta = \theta_1 - \theta_2 = 0.020$. Using Equation (7.36) and rounding up,

$$n \approx \left[(-1.645) + (-1.282)\right]^2 \left[(0.046)(0.954) + (0.026)(0.974)\right]\big/0.020^2 \approx 1{,}483.$$

7.97 Let X denote the number of persons who experience relief in a random sample of 30 persons treated with drug 1. Let Y denote the number of persons who experience relief in a random sample of 45 persons treated with drug 2. Let θ_1 and θ_2 denote the true proportions of persons treated with drug 1 and drug 2, respectively, who experience relief. We assume that $X \sim b(30, \theta_1)$ and $Y \sim b(45, \theta_2)$ are appropriate statistical models and that normal approximations are adequate.

$H_0: \theta_1 = \theta_2 \; ; H_a: \theta_1 > \theta_2$

Test Statistic: $Z = \left[\frac{X}{30} - \frac{Y}{45}\right]\big/\sqrt{\hat{\theta}(1 - \hat{\theta})\left[\frac{1}{30} + \frac{1}{45}\right]}$, with $\hat{\theta} = (X + Y)/75$

Decision Rule: Reject H_0 if p-value ≤ 0.10.

Analysis: Since $x = 18$ and $y = 23$, $z = \left[\frac{18}{30} - \frac{23}{45}\right]\big/\sqrt{\left(\frac{41}{75}\right)\left(\frac{34}{75}\right)\left[\frac{1}{30} + \frac{1}{45}\right]} \approx 0.76$ and the p-value is $P(Z \geq 0.76) = 0.22363$. There is insufficient evidence to reject the hypothesis that the two drugs are equally effective.

7.99 Let $\alpha = 0.10$, $Z_\alpha = -1.282$, $\beta = 0.10$, $Z_\beta = -1.282$, $\theta_1 = 0.60$, $\theta_2 = 0.51$, and $\delta = 0.05$. Using Equation (7.36) and rounding up,

$$n \approx \left[(-1.282) + (-1.282)\right]^2 \left[(0.60)(0.40) + (0.51)(0.49)\right]\big/0.05^2 \approx 1289.$$

7.101 $\left(\frac{309}{56,612} - \frac{27}{14,256}\right) - 1.645\sqrt{\left[\left(\frac{309}{56,612}\right)\left(\frac{56,303}{56,612}\right)\Big/56,612\right] + \left[\left(\frac{27}{14,256}\right)\left(\frac{14,229}{14,256}\right)\Big/14,256\right]} \approx 0.0028.$

We are 95% confident that the proportion of type A radios that will fail the customer check test exceeds that for the type B radios by at least 0.0028.

7.103 Let X denote the number of broken candy canes in a batch of 300 candy canes produced using the standard method. Let Y denote the number of broken candy canes in a batch of 200 candy canes produced using the new method. Let θ_1 and θ_2 denote the probabilities of producing a broken candy cane using the standard and new methods, respectively.

$H_0:\ \theta_1 = \theta_2\ ; H_a:\ \theta_1 > \theta_2$

Test Statistic: $\quad K = \sum_{k=0}^{Y} \frac{C(200, Y-k)C(300, X+k)}{C(500, X+Y)}$

Decision Rule: Letting PROB denote the observed value of K, reject H_0 if PROB $\leq \alpha$. (We will use $\alpha = 0.01$.)

Analysis: $x = 24$ and $y = 12$, so PROB $= \sum_{k=0}^{12} \frac{C(200, 12-k)C(300, 24+k)}{C(500, 36)} \approx$

0.2532. There is insufficient evidence to conclude that the proportion of broken candy canes produced using the new method is less than that for the standard method.

Section 7.6.1

7.105 (a) $H_0: \sigma^2 = 0.00156;\ H_a:\ \sigma^2 > 0.00156;$

Test Statistic: $W = 99S^2/0.00156;\ W \sim \chi^2_{(99)}$

Decision Rule: Reject H_0 if $w > 123.23$.
Analysis: Since $s^2 = 0.00211$, $w = 99(0.00211)/0.00156 \approx 133.90$. Therefore, we conclude that the new process is more variable than the old process.
Assumptions: The data were randomly selected from a normally distributed population.

(b) p-value $= P(W \geq 133.90) \approx 0.0112.$

7.107 (a) $H_0: \sigma^2 = 4;\ H_a:\ \sigma^2 \neq 4$

Test statistic: $W = 11S^2/4;\ W \sim \chi^2_{(11)}$

Decision rule: Since $W_{0.005} = 2.603$ and $W_{0.995} = 26.757$, reject H_0 if $w < 2.603$ or $w > 26.757$.
Analysis: Since $s \approx 3.84$, $w \approx 11(3.84^2)/4 = 40.5504$. We conclude that $\sigma^2 \neq 4$.

(b) $[11(3.84^2)/26.757,\ 11(3.84^2)/2.603] \approx [6.062, 62.313]$ —We are 99% confident that $6.062 \leq \sigma^2 \leq 62.313$.

(c) Assumptions: The data were randomly obtained from a normally distributed population.

(d) In the solution of Problem 7.47, the normal quantile plot is reasonably linear and the p-value of the Shapiro-Wilk W test is moderately large. Thus, a normality assumption is reasonable.

(e) Since a normality assumption is reasonable, the procedures used in (a) and (b) should provide reliable results.

7.109 (a) Since $\sigma_0 = \sqrt{0.000196} = 0.014$ and $\sigma = 0.024$, $\lambda = 0.024/0.014 \approx 1.7$. Using Figure L-5 of Appendix L, the intersection of the vertical line through $\lambda = 1.7$ and the $n = 5$ curve indicates that $\beta(1.7) \approx 0.50$.

(b) Using Figure L-5 of Appendix L, the intersection of the horizontal line through 0.10 on the $\beta(\lambda)$ axis and the vertical line through $\lambda = 1.5$ is very near the curve associated with $n = 30$. Therefore, a sample size of about 30 is required.

7.111 If $\sigma^2 = 0.000006$, $\sigma = \sqrt{0.000006} \approx 0.002449$. Since $\sigma_0 = \sqrt{0.000005} \approx 0.002236$, $\alpha = 0.05$, and an upper-tailed test is involved, we let $\lambda = 0.002449/0.002236 \approx 1.1$ and use Figure L-5 in Appendix L. A vertical line through $\lambda = 1.1$ intersects the curves associated with $n = 100$ and $n = 75$ near the point with ordinate $\beta = 0.60$. Since a sample of size 80 was used, $\beta(1.1) \approx 0.60$.

Section 7.6.2

7.113 H_0: $\sigma_X^2 = \sigma_Y^2$; H_a: $\sigma_X^2 > \sigma_Y^2$

Test statistic: $F = S_X^2 / S_Y^2$; $F \sim F_{(24,24)}$

Decision rule: Since $F_{0.95} = 1.984$, reject H_0 if $f > 1.984$.

Analysis: Since $s_X = \sqrt{30}$ and $s_Y = \sqrt{15}$, $f = 30/15 = 2$. Reject H_0 and conclude that $\sigma_X^2 > \sigma_Y^2$.

7.115 (a) H_0: $\sigma_A^2 = \sigma_B^2$; H_a: $\sigma_A^2 > \sigma_B^2$

Test statistic: $F = S_A^2 / S_B^2$; $F \sim F_{(20,20)}$

Decision rule: Since $F_{0.95} = 2.124$, reject H_0 if $f > 2.124$.

Analysis: Since $s_A^2 = 3,754,221$ and $s_B^2 = 1,473,659$, $f = \frac{3,7544,221}{1,473,659} \approx$ 2.548. Reject H_0 and conclude that the tensile strength of steel A is more variable than that of steel B.

(b) Let $\lambda = \sigma_A / \sigma_B = 2$ and use Figure L-7 in Appendix L. The ordinates of the intersections of the vertical line through $\lambda = 2$ and the curves associated with $n_A = n_B = 20$ and $n_A = n_B = 30$ are approximately 0.08 and 0.02, respectively. Since $n_A = n_B = 21$, $\beta(2) \approx 0.08$.

(c) The normal quantile plots displayed in the solution of Problem 6.55 are reasonably linear. The Shapiro-Wilk W test for

86

normality gives p-values of 0.7599 and 0.8255 for steels A and B, respectively. These test results and the normal quantile plots support our assumption that the two sampled populations are normally distributed.

7.117 Let $\lambda = 2$ and use Figure L-7 in Appendix L. The intersection of the vertical line through $\lambda = 2$ and the horizontal line through $\beta = 0.60$ falls between the curves associated with $n_A = n_B = 5$ and $n_A = n_B = 6$ Since that point is nearer the latter curve, random samples of size 6 should be obtained from each population.

7.118 To determine the common sample size needed to detect $\sigma_A = 1.5\sigma_B$ when $\alpha = 0.05$, let $\lambda = 1.5$ and use Figure L-7 in Appendix L. The intersection of the vertical line through $\lambda = 1.5$ and the horizontal line through $\beta = 0.05$ is close to the curve associated with $n_A = n_B = 75$. Thus, random samples of size 75 should be used.

7.119 Let X and Y denote the x-direction distances for the old and modified methods, respectively.

(a) $H_0: \sigma_X^2 = \sigma_Y^2; H_a: \sigma_X^2 > \sigma_Y^2$

Test statistic: $F = S_X^2 / S_Y^2 ; F \sim F_{(117, 122)}$
Decision rule: Reject H_0 if $f > 1.352$.
Analysis: Since $f = (1.397)^2 / (0.788)^2 \approx 3.143$, reject H_0 and conclude that $\sigma_X^2 > \sigma_Y^2$.

This solution is based on the assumptions that (1) the data were randomly obtained from two populations that are normally distributed and (2) those samples are independent.

(b) Since $F_{0.05} \approx 0.739$, $L = \frac{(1.397)^2}{(0.788)^2} \times 0.739 \approx 2.32$ mils^2 is a lower 95% confidence limit for σ_X^2 / σ_Y^2. We are 95% confident that $\sigma_X^2 \geq 2.32\,\sigma_Y^2$. This result is based on the same assumptions as those in part (a).

Section 7.7

7.121 Let X denote the distance a car travels on one gallon of gasoline and M denote the median of X.
$H_0: M = 54.0; H_a: M > 54.0$
Test statistic: The number, R, of times $X - 54$ is positive for a random sample of size 10. When H_0 is true, $R \sim b(10, 0.50)$.
Decision rule: Reject H_0 if p-value ≤ 0.05.
Analysis: Since 4 of the 10 data values exceed 54 and none equals 54, $r = 4$ for a sample of 10 non-zero values. $P[R \geq 4 \mid R \sim b(10, 0.50)] \approx 0.8281$ is the p-value. There is insufficient evidence to conclude that M > 54.0.

7.123 Let X denote the reading for a radio before vibration, Y denote the reading for that same radio after vibration, and $D = Y - X$. We assume that the set of differences is a random sample from a population of differences and that D is a random variable of the continuous type.
$H_0: M_D = 0; H_a: M_D > 0$
Test statistic: The number, R, of times D is positive for a random sample of size 10. When H_0 is true, $R \sim b(10, 0.50)$.
Decision rule: Reject H_0 if p-value $\leq \alpha$.
Analysis: The sample of differences is {3, 4, 1, 2, -1, -1, 3, 4, 5, 5}. Since 8 of these are positive, $r = 8$. The p-value is $P[R \geq 8 \mid R \sim b(10, 0.50)] \approx 0.0547$. Reject H_0 for any $\alpha \geq 0.0547$ and conclude that the distribution after vibration is shifted to the right of the distribution before vibration.

7.125 Assuming the distribution of the sampled population is symmetric and using the sign test, we test $H_0: \mu = 10.5$ versus $H_a: \mu > 10.5$. Since 7 of the 8 sample values are greater than 10.5 and 1 is less than 10.5, the p-value of the test is $P[R \geq 7 \mid R \sim b(8, 0.5)] = 0.0352$. But $\alpha = 0.10$, so we reject H_0 and conclude that (on the average) more than 10.5 days are required to fill the orders.

7.127 Since a normal distribution is symmetric, the Wilcoxon signed rank test can be used.
$H_0: \mu = 1.11; H_a: \mu > 1.11$
Test statistic: $W = T_-$
Decision rule: Reject H_0 if p-value ≤ 0.01.
Analysis: Using the summary table that follows, $w = t_- = \mid -1.5 \mid = 1.5$. Since $P(W \leq 1) = 0.05$ when a sample of 5 nonzero deviations is considered, the p-value is greater than 0.05. Therefore, there is insufficient evidence to reject H_0.

i	1	2	3	4	5
x_i	1.14	1.15	1.14	1.12	1.10
$x_i - 1.11$	0.03	0.04	0.03	0.01	-0.01
$\mid x_i - 1.11 \mid$	0.03	0.04	0.03	0.01	0.01
r_i	3.5	5	3.5	1.5	1.5
s_i	3.5	5	3.5	1.5	-1.5

7.129 We assume symmetry and use the Wilcoxon signed rank test.
$H_0: \mu = 0.85; H_a: \mu \neq 0.85$
Test statistic: Z, as defined in Equation (7.41), will be used.
Decision rule: Reject H_0 if p-value ≤ 0.01.
Analysis: Choosing the **Wilcoxon signed-ranks** option in the **Nonparametric tests** pop-up menu of the **Statistics** module in StataQuest, an observed value of $z = -0.15$ is obtained. Since $P(Z \leq -0.15) = 0.44038$, the p-value is $2(0.44038) = 0.88076$. (*Note:* StataQuest calculates the p-value before rounding z to two decimal places and reports 0.8774.) There is insufficient evidence to reject H_0.

7.131 Let X and Y denote the amounts of creosote extracted by solvents A and B, respectively. Letting $D = Y - X$ and assuming that D has a symmetric distribution, we use the Wilcoxon signed rank test. Let M denote the median of the distribution of D.

$H_0: M = 0$; $H_a: M > 0$
Test statistic: $W = T_-$
Decision rule: Reject H_0 if p-value $\leq \alpha$, where $\alpha = 0.05$.
Analysis: Using the summary table that follows, $w = t_- = |{-3} + {-2}| = 5$. Since $P(W \leq 4) = 0.05$ when a sample of 7 nonzero deviations is considered, the p-value is greater than 0.05. Therefore, do not reject H_0.

i	1	2	3	4	5	6	7
d_i	0.52	0.41	0.75	-0.20	0.81	-0.19	0.14
$d_i - 0$	0.52	0.41	0.75	-0.20	0.81	-0.19	0.14
$\lvert d_i - 0 \rvert$	0.52	0.41	0.75	0.20	0.81	0.19	0.14
r_i	5	4	6	3	7	2	1
s_i	5	4	6	-3	7	-2	1

7.133 We assume that independent random samples were selected.

$H_0: M_1 = M_2$; $H_a: M_1 \neq M_2$
Test statistic: $T = T_1$
Decision rule: For $\alpha = 0.05$, reject H_0 if $t \leq 29$ or $t \geq 61$.
Analysis: The ranks of the plant 1 tensile strengths are 5, 2, 6, 9.5, 3, and 8. Thus, $t = 5 + 2 + 6 + 9.5 + 3 + 8 = 33.5$. There is insufficient evidence to conclude that the median tensile strengths differ.

7.135 **(a)** Using MYSTAT, the following grouped, notched box plot display was obtained. Since the notches do not overlap, we can reject equality of medians at an approximate 5% significance level.

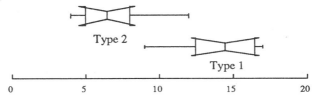

(b) We assume that independent random samples were selected.

$H_0: M_1 = M_2$; $H_a: M_1 \neq M_2$
Test statistic: $T = T_2$
Decision rule: For $\alpha = 0.05$, reject H_0 if $t \leq 29$ or $t \geq 61$.
Analysis: The ranks of the type 2 shrinks are 1, 2, 3, 4, 5, and 7.5. Thus, $t = 1 + 2 + 3 + 4 + 5 + 7.5 = 22.5$. Reject H_0 and conclude that the median shrinks for the two molds are not equal.

7.137

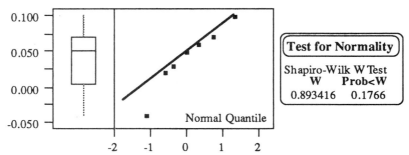

Since a pair of switches was measured for each of 10 crash sensors, we have paired observations. Letting $D = B - A$, the observed differences are 0.05, 0.02, 0.06, 0.07, 0.03, -0.04, 0.05, 0.07, 0.10, and -0.04.

Even though the p-value of the Shapiro-Wilk W Test for these differences is not very small (0.1766), the normal quantile plot gives the impression that the data may have been obtained from a population with a skewed distribution. Thus, we will use the one-sample sign test to test $H_0: M_D = 0$ versus $H_a: M_D \neq 0$.

Since 8 of the 10 observed differences are positive and 2 are negative, the p-value of the test is $2P[R \leq 2 \mid R \sim b(10, 0.5)] = 0.1094$. We can reject H_0 for any $\alpha \geq 0.1094$. Thus, we doubt that the median closure velocities of the two types of switches differ.

CHAPTER 8

SHEWHART CONTROL CHARTS

Sections 8.1.1 and 8.1.2

8.1   If the range is not in control, a valid estimate of a population variance cannot be obtained.

8.3   **(a)** $\mu_X \approx \bar{\bar{x}} = 14.300$ and $\sigma_X \approx \bar{R}/d_2 = 0.352/2.326 \approx 0.151$

$$P(14.000 \leq X \leq 14.800) \approx P\left(\tfrac{14.000-14.300}{0.151} \leq Z \leq \tfrac{14.800-14.300}{0.151}\right)$$
$$\approx P(-1.99 \leq Z \leq 3.31)$$
$$= 0.99953 - 0.02330 = 0.97623$$

Approximately 97.6% of the production is expected to fall within the specification limits.

(b) Reducing variability would increase the percent within specifications, if centering remained at 14.300. Also, if centered at 14.40 and $\sigma_X = 0.151$,

$$P(14.000 \leq X \leq 14.800) = P\left(\tfrac{14.000-14.400}{0.151} \leq Z \leq \tfrac{14.800-14.400}{0.151}\right)$$
$$\approx P(-2.65 \leq Z \leq 2.65)$$
$$= 0.99598 - 0.00402 = 0.99196,$$

and about 99.2% of the production would be within the specification limits.

8.5   R chart: $\bar{R} = (23 + 8 + \ldots + 3 + 8)/20 = 191/20 = 9.55$, $D_3 = 0$, $D_4 = 2.115$, LCL $= D_3\bar{R} = 0.00$, CL $= 9.55$, and UCL $= D_4\bar{R} = (2.115)(9.55) \approx 20.20$

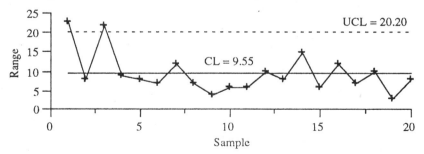

$\bar{x}$ chart: $\bar{\bar{x}} = (7.6 + 6.6 + \ldots + 9.0 + 11.1)/20 = 172.7/20 = 8.635$,

$A_2 = 0.577$, $\bar{R} \approx 9.55$, LCL $= \bar{\bar{x}} - A_2\bar{R} = 8.635 - (0.577)(9.55) \approx 3.12$,

CL $= 8.635 \approx 8.64$, and UCL $= \bar{\bar{x}} + A_2\bar{R} = 8.635 + (0.577)(9.55) \approx 14.15$

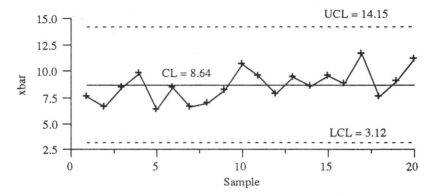

Consider the preceding JMP range chart. The ranges for samples 1 and 3 are beyond the upper control limit for R, indicating out-of-control conditions. The cyclic pattern for samples 11 through 20 may also be a signal of an out-of-control condition.

8.7 The correct answer is (d).

8.9 A summary of the sample means and ranges follows. Sample means have been rounded to the nearest hundredth.

Sample	1	2	3	4	5	6	7	8	9	10
$\bar{x}$	4.30	4.07	4.18	4.05	4.13	4.10	4.38	3.95	3.87	3.65
R	1.4	0.4	1.4	0.9	1.0	1.4	1.4	1.3	1.1	1.2

Sample	11	12	13	14	15	16	17	18	19	20
$\bar{x}$	3.90	3.90	3.57	3.87	3.77	3.55	3.88	4.00	4.38	4.22
R	1.0	1.1	0.7	0.6	1.3	0.8	0.6	0.4	1.0	0.6

R chart: $\bar{R} = (1.4 + \ldots + 0.6)/20 = 19.6/20 = 0.98$, $D_3 = 0$, $D_4 = 2.004$, LCL $= D_3\bar{R} = 0.00$, CL $= 0.98$, and UCL $= D_4\bar{R} = (2.004)(0.98) \approx 1.96$

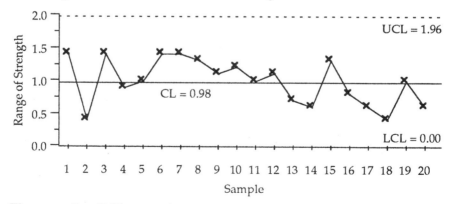

The preceding JMP range chart contains no points beyond the control limits. The pattern of the points does not appear unusual.

$\bar{x}$ chart: $\overline{\overline{x}} \approx (4.30 + \ldots + 4.22)/20 = 79.72/20 \approx 3.99$, $A_2 = 0.483$,
$\overline{R} = 0.98$, LCL $= \overline{\overline{x}} - A_2\overline{R} = 3.99 - (0.483)(0.98) \approx 3.52$, CL $= 3.99$, and
UCL $= \overline{\overline{x}} + A_2\overline{R} = 3.99 + (0.483)(0.98) \approx 4.46$

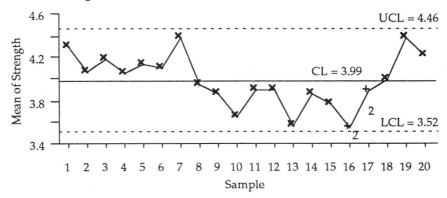

The first 7 points on the $\bar{x}$ chart are above the center line, and the next
10 points are below that line. This seems to indicate a shift in the
process centering. JMP indicates that 9 consecutive points are on the
same side of the center line by affixing a 2 to the 9th point. This
indicates that the process is out of control (Section 8.2.2).

8.11

Sample	1	2	3	4	5	6	7	8
$\bar{x}$	25.0	10.0	22.2	26.0	2.4	3.0	2.8	17.0
R	34	14	29	23	3	4	2	18

Sample	9	10	11	12	13	14	15	16
$\bar{x}$	19.6	14.4	12.0	11.2	7.4	9.4	11.4	10.6
R	24	18	21	9	9	14	9	16

Sample	17	18	19	20	21	22	23	24
$\bar{x}$	23.8	15.0	12.8	13.2	13.0	13.4	16.4	17.4
R	35	20	25	19	12	14	13	22

R chart: $\overline{R} = (34 + \ldots + 22)/24 = 407/24 \approx 16.96$, $D_3 = 0$, $D_4 = 2.115$,
LCL $= D_3\overline{R} = 0.0$, CL $= \overline{R} \approx 17.0$, and UCL $= D_4\overline{R} = (2.115)(16.96) \approx 35.9$

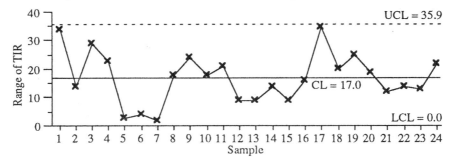

93

The preceding JMP range chart contains no points beyond the control limits. The process range appears to be in control.

$\bar{x}$ chart: $\bar{\bar{x}} \approx (25.0 + \ldots + 17.4)/24 = 329.4/24 = 13.725 \approx 13.73$,
$A_2 = 0.577$, $\bar{R} \approx 16.96$, LCL $= \bar{\bar{x}} - A_2\bar{R} = 13.73 - (0.577)(16.96) \approx 3.9$,
CL $= \bar{\bar{x}} \approx 13.7$, and UCL $= \bar{\bar{x}} + A_2\bar{R} = 13.73 + (0.577)(16.96) \approx 23.5$

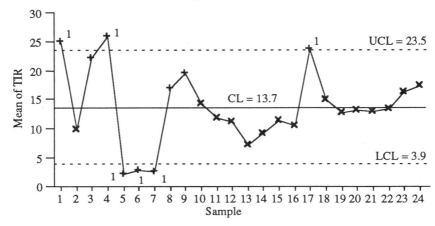

JMP indicates that the points associated with samples 1, 4, 5, 6, 7, and 17 are outside the control limits for $\bar{x}$ by affixing the numeral 1 beside each. The process mean is not in control.

8.13

Sample	1	2	3	4	5	6	7	8	9	10
$\bar{x}$	10.8	11.6	10.8	9.0	8.8	12.4	9.8	10.8	11.2	10.2
R	7	9	6	6	10	9	7	12	5	8

Sample	11	12	13	14	15	16	17	18	19	20
$\bar{x}$	10.8	9.0	9.0	10.2	12.0	9.0	12.4	10.2	15.0	9.4
R	6	9	8	9	11	7	18	8	4	13

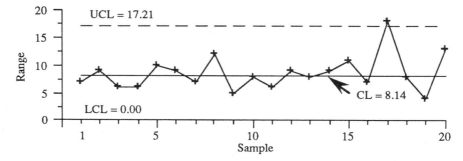

The range of sample 17 exceeds the upper control limit for R, so the standard of $\sigma = 3.5$ is not being met. Also, the mean of sample 19 exceeds the upper control limit for $\bar{x}$, so the standard of $\mu = 10$ is not being met.

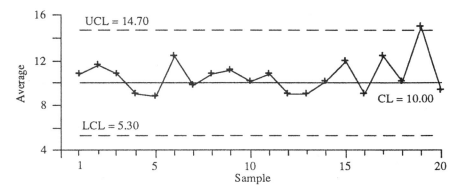

8.15 In Problem 8.13, the R chart is used to test H_{01}: "The true population standard deviation is 3.5." Since the range of sample 17 exceeds the upper control limit, H_0 is rejected.

In Problem 8.14, the R chart is used to test H_{02}: "The samples were obtained from a population with constant, but unknown, standard deviation." There is insufficient evidence to reject this hypothesis.

In Problem 8.13, the $\bar{x}$ and R charts are used together to test H_{03}: "The samples were obtained from a population with mean 10 and standard deviation 3.5." Since the range is not in control, we reject this hypothesis.

In Problem 8.14, the $\bar{x}$ and R charts are used together to test H_{04}: "The samples were obtained from an unspecified but constant population." There is insufficient evidence to reject this hypothesis.

8.17

Sample	1	2	3	4	5	6	7	8	9	10
$\bar{x}$	4.5	4.6	4.2	4.1	4.1	4.2	4.3	4.2	4.5	4.8
R	1.0	0.5	0.5	0.5	0.5	0.5	0.5	0.5	1.0	1.5

Sample	11	12	13	14	15	16	17	18	19	20
$\bar{x}$	4.5	4.9	4.8	4.2	4.9	4.2	3.6	4.3	4.5	4.6
R	0.0	1.5	1.0	1.0	1.5	0.5	1.0	0.5	0.0	0.5

R chart: $\bar{R} = (1.0 + \ldots + 0.5)/20 = 14.5/20 = 0.725$, $D_3 = 0$, $D_4 = 2.115$, LCL $= D_3\bar{R} = 0.00$, CL $= \bar{R} \approx 0.73$, and UCL $= D_4\bar{R} \approx (2.115)(0.725) \approx 1.53$

$\bar{x}$ chart: $\bar{\bar{x}} = (4.5 + \ldots + 4.6)/20 = 88.0/20 = 4.40$, $A_2 = 0.577$, $\bar{R} = 0.725$, LCL $= \bar{\bar{x}} - A_2\bar{R} \approx 4.40 - (0.577)(0.725) \approx 3.982$, CL $= \bar{\bar{x}} = 4.400$, and UCL $= \bar{\bar{x}} + A_2\bar{R} \approx 4.40 + (0.577)(0.725) \approx 4.818$

From the JMP range chart that follows, the range appears to be in control. The averages associated with samples 12 and 15 on the following $\bar{x}$ chart are greater than UCL; that for sample 17 is less than LCL. The process centering is not stable.

95

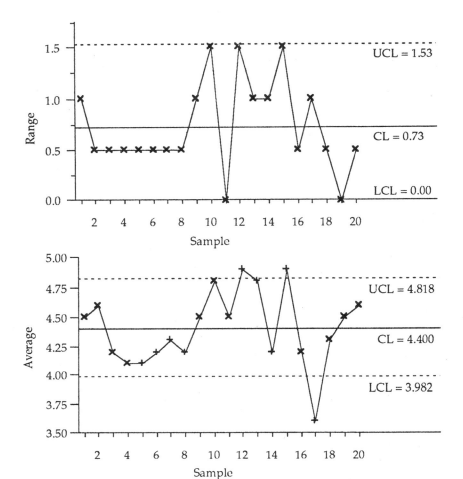

8.19

Sample	1	2	3	4	5	6	7	8	9
$\bar{x}$	2.92	4.80	3.54	2.86	2.44	3.24	3.88	2.06	5.22
R	3.40	3.90	2.50	1.50	3.80	4.10	2.60	1.70	6.20

Sample	10	11	12	13	14	15	16	17	18
$\bar{x}$	3.66	4.70	3.50	3.24	4.84	2.70	3.84	2.84	2.30
R	4.10	8.00	4.90	1.50	2.10	2.70	2.10	3.60	2.80

Sample	19	20	21	22	23	24	25		
$\bar{x}$	4.04	2.90	1.82	3.58	2.78	4.02	2.22		
R	3.50	2.10	2.20	4.10	3.00	3.70	2.10		

R chart: $\bar{R} = (3.40 + \ldots + 2.10)/25 = 82.2/25 = 3.288$, $D_3 = 0$, $D_4 = 2.115$,
$LCL = D_3\bar{R} = 0.0$, $CL = \bar{R} \approx 3.3$, and $UCL = D_4\bar{R} \approx (2.115)(3.288) \approx 7.0$

For the following R chart, the range of sample 11 exceeds the upper
control limit. Thus, R is not in control.

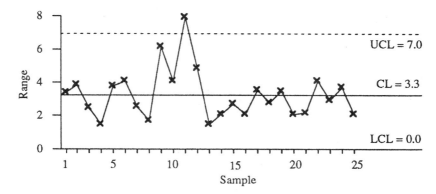

$\bar{x}$ chart: $\bar{\bar{x}} = (2.92 + \ldots + 2.22)/25 = 83.94/25 = 3.3576$, $A_2 = 0.577$, $\bar{R} = 3.288$, $\text{LCL} = \bar{\bar{x}} - A_2\bar{R} = 3.3576 - (0.577)(3.288) \approx 1.46$, $\text{CL} = \bar{\bar{x}} \approx 3.36$, and $\text{UCL} = \bar{\bar{x}} + A_2\bar{R} \approx 3.3576 + (0.577)(3.288) \approx 5.25$

Use of the $\bar{x}$ chart is questionable when R is not in control. Based on the text discussion to this point, the following control chart seems to indicate that the process centering is stable.

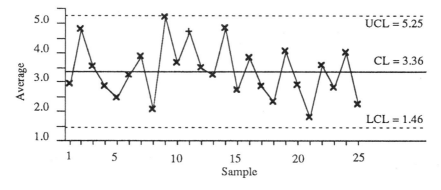

8.21 For samples of size 4 , $d_2 = 2.059$; so, $\hat{\sigma}_X = \bar{R}_1/d_2 = 0.040/2.059$. For samples of size 9, $d_2 = 2.970$; so, $\hat{\sigma}_Y = \bar{R}_2/d_2 = 0.005/2.970$. Thus, $Var(W) \approx (0.040/2.059)^2 + (0.005/2.970)^2 \approx 0.00038$ and $\sigma_w \approx \sqrt{0.00038} \approx 0.0195$. Since $\mu_w = \mu_X + \mu_Y = 8.5 + 6.5 = 15.0$, $P(W > 15.006) \approx P[Z > (15.006 - 15.000)/0.0195)] \approx P(Z > 0.31) = 0.37828$.

Sections 8.1.4 and 8.1.5

8.23 Control limits are based on standards of $\mu = 10$ and $\sigma = 3.5$.

s **chart:** $\sigma = 3.5$, $B_5 = 0$, $B_6 = 1.964$, $c_4 = 0.940$, $\text{LCL} = B_5\sigma = 0.00$, $\text{CL} = c_4\sigma = (0.940)(3.5) \approx 3.29$, and $\text{UCL} = B_6\sigma = (1.964)(3.5) \approx 6.87$

$\bar{x}$ chart: $\mu = 10$, $\sigma = 3.5$, $A = 1.342$, LCL $= \mu - A\sigma = 10 - (1.342)(3.5) \approx 5.30$, CL $= \mu = 10$, and UCL $= \mu + A\sigma = 10 - (1.342)(3.5) \approx 14.70$

Using the data in Problem 8.13, we obtain the following summary of sample means and standard deviations (calculated to the nearest 0.01).

Sample	1	2	3	4	5	6	7	8	9	10
$\bar{x}$	10.80	11.60	10.80	9.00	8.80	12.40	9.80	10.80	11.20	10.20
s	2.77	3.78	2.39	2.45	4.32	3.91	3.42	4.82	1.92	3.27
Sample	11	12	13	14	15	16	17	18	19	20
$\bar{x}$	10.80	9.00	9.00	10.20	12.00	9.00	12.40	10.20	15.00	9.40
s	2.28	3.32	2.92	3.42	3.94	2.74	6.88	3.35	1.58	5.32

The calculated control limits and points summarized in the preceding table are given on the following control charts. Since the standard deviation of sample 17 exceeds the upper control limit for s, the standard of $\sigma = 3.5$ is not being met. The average of sample 19 exceeds the upper control limit for $\bar{x}$, so the standard of $\mu = 10$ is not being met.

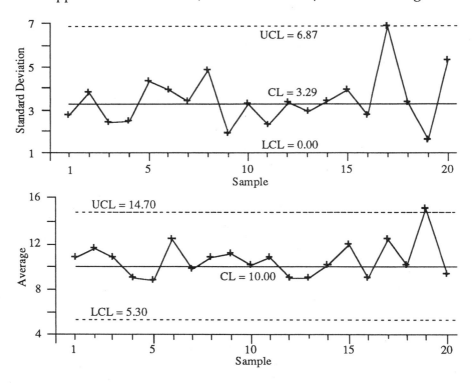

8.25   In Problem 8.23, the s chart is used to test H_{01}: "The true population standard deviation is 3.5." Since the standard deviation of sample 17 exceeds the upper control limit, this hypothesis is rejected.

In Problem 8.24, the s chart is used to test H_{02}: "The samples were obtained from a population with constant, but unknown, standard deviation." There is insufficient evidence to reject this hypothesis.

In Problem 8.23, the $\bar{x}$ and s charts are used together to test H_{03}: "The samples were obtained from a population with mean 10 and standard deviation 3.5." Since the standard deviation is not in control, we reject this hypothesis.

In Problem 8.24, the $\bar{x}$ and s charts are used together to test H_{04}: "The samples were obtained from an unspecified but constant population." There is insufficient evidence to reject this hypothesis.

8.27

Sample	1	2	3	4	5	6	7	8	9	10
$\bar{x}$	104.0	76.0	77.8	85.0	84.0	79.2	79.8	86.0	70.8	80.8
s	1.6	13.7	4.1	2.9	10.4	3.8	3.6	2.4	3.3	3.3

Sample	11	12	13	14	15	16	17	18	19	20
$\bar{x}$	81.0	74.2	74.5	102.8	106.2	89.5	104.2	96.8	100.2	93.2
s	1.8	4.1	6.6	6.8	5.7	3.9	4.9	5.4	4.6	4.9

(a) s chart: $\bar{s} = (1.6 + \ldots + 4.9)/20 = 97.8/20 = 4.89$, $B_3 = 0$, $B_4 = 2.266$, LCL $= B_3\bar{s} = 0.00$, CL $= \bar{s} = 4.89$, and UCL $= B_4\bar{s} \approx (2.266)(4.89) \approx$ 11.08. As shown on the following s chart, the standard deviation of sample 2 exceeds UCL. This may indicate the presence of a special cause of variation.

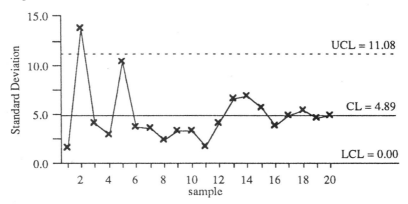

$\bar{x}$ chart: $\bar{\bar{x}} = (104.0 + \ldots + 93.2)/20 = 1746.0/20 = 87.30$, $A_3 = 1.628$, $\bar{s} = 4.89$, LCL $= \bar{\bar{x}} - A_3\bar{s} = 87.30 - (1.628)(4.89) \approx 79.34$, CL $= \bar{\bar{x}} =$ 87.30, and UCL $= \bar{\bar{x}} + A_3\bar{s} \approx 87.30 + (1.628)(4.89) \approx 95.26$. On the following $\bar{x}$ chart, many sample means are outside the control limits. It appears that a shift in the process centering occurred about the time that sample 14 was obtained.

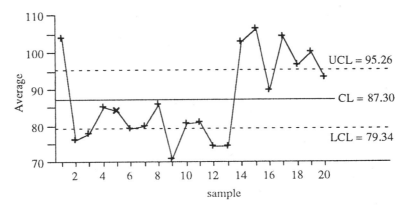

(b) Since s is not in control, an estimate of a common σ cannot be made.

Section 8.2

8.29 (a) No. A point is outside the control limits.
    (b) No. The first nine consecutive points are above the center line.
    (c) No. The first nine points in a row continue to increase.
    (d) No. Consecutive points are alternating up and down.

8.31 (a) A special cause produces variation that cannot be explained by a single distribution.
    (b) A common cause produces variation that is the responsibility of management to correct.

8.33 Small, rational subgroups are chosen to minimize the chances of special causes occurring within the sample while maximizing the chances of detecting special causes that occur among the samples.

Section 8.3

8.35 (a)

Device	1	2	3	4	5
x	4290	4100	4210	4080	4350
R	———	190	110	130	270
Device	6	7	8	9	10
x	4185	4320	4050	4230	4160
R	165	135	270	180	70

$\overline{R} = \frac{190 + \ldots + 70}{9} = \frac{1520}{9} \approx 168.9$, $\overline{x} = \frac{4290 + \ldots + 4160}{10} = \frac{41{,}975}{10} = 4197.5$,
LCL = $\overline{x}$ - 2.660 $\overline{R}$ $\approx$ 4197.5 - (2.660)(168.9) $\approx$ 3748.2, CL = $\overline{x}$ = 4197.5, and UCL = $\overline{x}$ + 2.660 $\overline{R}$ $\approx$ 4197.5 + (2.660)(168.9) $\approx$ 4646.8.

All points in the following chart are within the control limits and no unusual trends or runs are present.

100

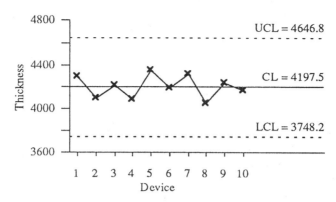

(b) The two scatter plots are equivalent. The digidot plot includes a stem-and-leaf display, indicating that an assumption of normality is reasonable. The control limits on the individuals charts makes tests for unusual patterns and trends possible.

Section 8.4.1

8.37

Sample	1	2	3	4	5	6	7	8
x	11	13	4	9	14	7	6	12
n	128	151	105	137	143	112	98	138
p	0.086	0.086	0.038	0.066	0.098	0.062	0.061	0.087

Sample	9	10	11	12	13	14	15
x	9	10	10	11	8	11	13
n	123	119	130	141	129	136	142
p	0.073	0.084	0.077	0.078	0.062	0.081	0.092

$CL = \bar{p} = \sum x_i / \sum n_i = 148/1932 \approx 0.077$; $\bar{n} = \sum n_i/15 = 1932/15 = 128.8$;

$LCL \approx \bar{p} - 3\sqrt{\bar{p}(1-\bar{p})/\bar{n}} = 0.077 - 3\sqrt{(0.077)(0.923)/128.8} \approx 0.077 - 0.070 = 0.007$; and $UCL \approx \bar{p} + 3\sqrt{\bar{p}(1-\bar{p})/\bar{n}} = 0.077 + 3\sqrt{(0.077)(0.923)/128.8} \approx 0.077 + 0.070 = 0.147$. All points are within their control limits. No unusual trends or patterns are present in the following control chart. The population proportion appears to be in control.

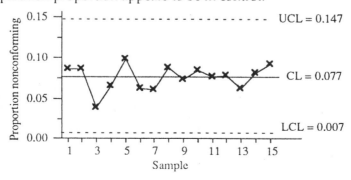

8.39

Sample	1	2	3	4	5	6	7
x	9	15	7	10	6	11	13
n	40	40	40	40	40	40	40
p	0.225	0.375	0.175	0.250	0.150	0.275	0.325
Sample	8	9	10	11	12	13	14
x	8	10	12	5	9	14	11
n	40	40	40	40	40	40	40
p	0.200	0.250	0.300	0.125	0.225	0.350	0.275

$CL = \bar{p} = (9 + 15 + \ldots + 14 + 11)/(40)(14) = 0.250$

$LCL = 0.250 - 3\sqrt{(0.250)(0.750)/40} \approx 0.045$

$UCL = 0.250 + 3\sqrt{(0.250)(0.750)/40} \approx 0.455.$

The following control chart gives no indication of the presence of special causes.

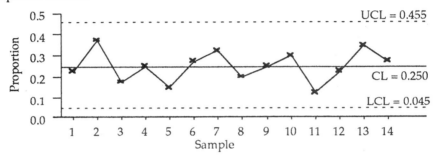

8.41 $CL = \theta = 0.060$ and $UCL = \theta + 3\sqrt{\theta(1 - \theta)/n} = 0.06 + 3\sqrt{(0.06)(0.94)/25} \approx 0.202 \approx 0.20$. Since $\theta - 3\sqrt{\theta(1 - \theta)/n} = 0.06 - 3\sqrt{(0.06)(0.94)/25}$ is negative, let $LCL = 0.000$.

8.43

Sample	1	2	3	4	5	6	7	8	9	10
x	1	2	2	0	1	0	2	2	0	3
p	0.04	0.08	0.08	0.00	0.04	0.00	0.08	0.08	0.00	0.12
Sample	11	12	13	14	15	16	17	18	19	20
x	1	1	0	1	6	2	1	0	1	0
p	0.04	0.04	0.00	0.04	0.24	0.08	0.04	0.00	0.04	0.00

$CL = \bar{p} = (1 + 2 + \ldots + 1 + 0)/(25)(20) = 0.052$

$LCL = 0$ because $0.052 - 3\sqrt{(0.052)(0.948)/25}$ is negative

$UCL = 0.052 + 3\sqrt{(0.052)(0.948)/25} \approx 0.185.$

On the following control chart, the proportion of failures for sample 15 exceeds UCL. Conditions at the time sample 15 was obtained should be carefully studied to see if a reason for this unusual occurrence can be identified. The chart presents no other signal of the presence of special cause variation.

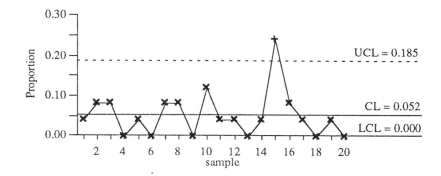

8.45

Sample	1	2	3	4	5	6	7	8
x	5	11	3	2	8	9	5	12
n	180	212	148	165	201	173	188	192
p	0.028	0.052	0.020	0.012	0.040	0.052	0.027	0.062

Sample	9	10	11	12	13	14	15	16
x	7	6	10	4	7	9	4	14
n	193	169	183	171	197	188	195	211
p	0.036	0.036	0.055	0.023	0.036	0.048	0.021	0.066

$CL = \bar{p} = \sum x_i / \sum n_i = \frac{116}{2966} \approx 0.039$; $\bar{n} = \sum n_i / 16 = \frac{2966}{16} = 185.375$

$UCL \approx \bar{p} + 3\sqrt{\bar{p}(1-\bar{p})/\bar{n}} = 0.039 + 3\sqrt{\frac{(0.039)(0.961)}{185.375}} \approx 0.039 + 0.043 = 0.082.$

Since $\bar{p} - 3\sqrt{\bar{p}(1-\bar{p})/\bar{n}} \approx 0.039 - 0.043$ is negative, let LCL = 0.000.

All points are within their control limits. No unusual trends or patterns are present in the following control chart. The population proportion appears to be in control.

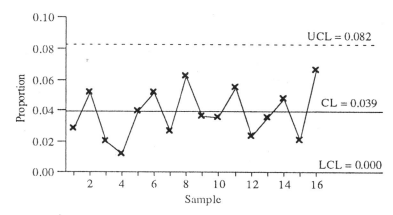

8.47 **(a)** Let x_i denote the number of loose nuts in the ith sample of 800 nuts. Since $\sum x_i = (272 + \ldots + 188) = 5167$ for the $24(800) = 19{,}200$ nuts, $CL = \bar{p} = 5167/19{,}200 \approx 0.269$; $LCL = 0.269 - 3\sqrt{(0.269)(0.731)/800} \approx 0.269 - 0.047 = 0.222$; and $UCL \approx 0.269 + 0.047 = 0.316$.

(b)

Sample	1	2	3	4	5	6	7	8
x	272	274	234	165	206	203	221	201
n	800	800	800	800	800	800	800	800
p	0.340	0.342	0.292	0.206	0.258	0.254	0.276	0.251

Sample	9	10	11	12	13	14	15	16
x	208	197	221	232	210	255	288	295
n	800	800	800	800	800	800	800	800
p	0.260	0.246	0.276	0.290	0.262	0.319	0.360	0.369

Sample	17	18	19	20	21	22	23	24
x	155	187	158	207	219	195	176	188
n	800	800	800	800	800	800	800	800
p	0.194	0.234	0.198	0.259	0.274	0.244	0.220	0.235

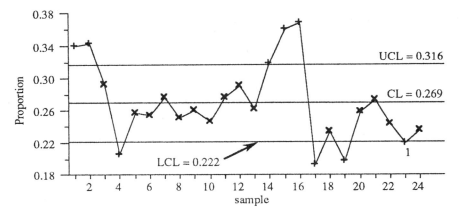

The proportions associated with samples 1, 2, 14, 15, and 16 exceed UCL. The operators who assembled the bumpers in those samples were not the regular operators. It appears that those new operators need more training.

The proportions associated with samples 4, 17, 19, and 23 are less than LCL. These samples may signal cases for which process improvement has occurred.

8.49 **(a)** No. The average sample size is $\bar{n} = \sum n_i/20 = 232{,}400/20 = 11{,}620$. The only sample sizes contained in the interval $[0.8\,\bar{n}, 1.2\,\bar{n}] = [9{,}296, \ 13{,}944]$ are those for samples 7, 9, 12, and 15.

(b)

Sample	x	n	p	LCL	UCL
1	61	24,960	0.002	0.005	0.008
2	22	960	0.023	0.000	0.014
3	24	2,880	0.008	0.002	0.011
4	10	1,920	0.005	0.001	0.012
5	29	3,840	0.008	0.002	0.010
6	4	3,960	0.001	0.003	0.010
7	214	13,320	0.016	0.004	0.008
8	25	14,400	0.002	0.004	0.008
9	150	11,520	0.013	0.004	0.009
10	417	37,184	0.011	0.005	0.008
11	6	5,760	0.001	0.003	0.010
12	42	11,520	0.004	0.004	0.009
13	82	8,640	0.009	0.004	0.009
14	84	16,320	0.005	0.004	0.008
15	48	10,560	0.005	0.004	0.009
16	21	7,680	0.003	0.004	0.009
17	23	16,320	0.001	0.004	0.008
18	90	16,320	0.006	0.004	0.008
19	16	5,760	0.003	0.003	0.010
20	119	18,576	0.006	0.005	0.008
Total	1,487	232,400			

Since $\sum x_i = 61 + \ldots + 119 = 1487$ for the $\sum n_i = 24{,}960 + \ldots + 18{,}576 = 232{,}400$ circuit boards, CL $= \bar{p} = 1487/232{,}400 \approx 0.0064$. Control limits for the ith sample (given in the preceding table) are calculated using

$$\text{LCL} = 0.0064 - 3\sqrt{(0.0064)(0.9936)/n_i}$$

and

$$\text{UCL} = 0.0064 + 3\sqrt{(0.0064)(0.9936)/n_i}.$$

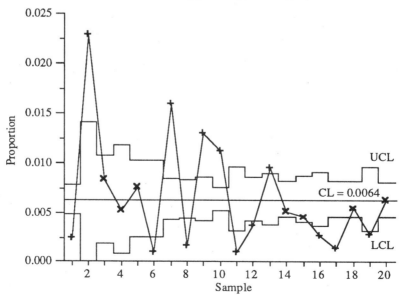

Using the first 3 columns of the preceding table, JMP produced the preceding p chart. The sample 2, 7, 9, 10, and 13 proportions exceed their upper control limits. The circumstances at the time those samples were obtained should be investigated fully.

The proportions associated with samples 1, 6, 8, 11, 12, 16, 17, and 19 are less than their lower control limits. Careful study of the circumstances at the time those samples were obtained may provide clues for process improvement.

8.51 **(a)** $LCL = \theta - 3\sqrt{\theta(1-\theta)/n} = 0.120 - 3\sqrt{(0.120)(0.880)/600} \approx 0.120 - 0.040$
$= 0.080$ and $UCL = \theta + 3\sqrt{\theta(1-\theta)/n} \approx 0.120 + 0.040 = 0.160$.

(b) None of the proportions falls outside the control limits.

8.53 **(a)** $LCL = \theta - 3\sqrt{\theta(1-\theta)/n} = 0.220 - 3\sqrt{(0.220)(0.780)/600} \approx 0.220 - 0.051$
$= 0.169$ and $UCL = \theta + 3\sqrt{\theta(1-\theta)/n} \approx 0.220 + 0.051 = 0.271$.

(b) None of the proportions falls outside the control limits.

8.55 **(a)** $LCL = \theta - 3\sqrt{\theta(1-\theta)/n} = 0.160 - 3\sqrt{(0.160)(0.840)/600} \approx 0.160 - 0.045$
$= 0.115$ and $UCL = \theta + 3\sqrt{\theta(1-\theta)/n} \approx 0.160 + 0.045 = 0.205$.

(b) None of the proportions falls outside the control limits.

Section 8.4.2

8.57

Sample	1	2	3	4	5	6	7	8	9	10
c	21	23	18	22	26	25	17	24	30	27
a	10	10	10	10	10	10	10	10	10	10
u	2.1	2.3	1.8	2.2	2.6	2.5	1.7	2.4	3	2.7

Sample	11	12	13	14	15	16	17	18	19	20
c	32	21	28	21	30	29	28	20	23	22
a	10	10	10	10	10	10	10	10	10	10
u	3.2	2.1	2.8	2.1	3	2.9	2.8	2	2.3	2.2

(a) $\bar{u} = \Sigma c_i / \Sigma a_i = (21 + \ldots + 22)/(10 + \ldots + 10) = 487/200 = 2.435$;

$LCL = \bar{u} - 3\sqrt{\bar{u}/a} = 2.435 - 3\sqrt{2.435/10} \approx 0.95$; $CL = \bar{u} \approx 2.44$; and

$UCL = \bar{u} + 3\sqrt{\bar{u}/a} = 2.435 + 3\sqrt{2.435/10} \approx 3.92$.

(b)

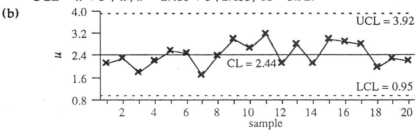

The JMP u chart contains no signal of an out-of-control condition.

8.59 $\bar{u} = \Sigma c_i / \Sigma a_i = 5235/1073 \approx 4.88$. For the ith sample, the standardized value of u_i is $z_i = (u_i - 4.88)\sqrt{a_i}/\sqrt{4.88}$. Using this formula with the data in Problem 8.58 gives the following table.

Sample	1	2	3	4	5	6	7
c	204	321	192	297	253	341	409
a	63	95	48	72	67	71	83
u	3.24	3.38	4.00	4.12	3.78	4.80	4.93
z	−5.90	−6.62	−2.76	−2.92	−4.08	−0.30	0.21

Sample	8	9	10	11	12	13	14
c	401	373	511	465	451	527	490
a	79	65	96	88	74	90	82
u	5.08	5.74	5.32	5.28	6.09	5.86	5.98
z	0.80	3.14	1.95	1.70	4.71	4.21	4.51

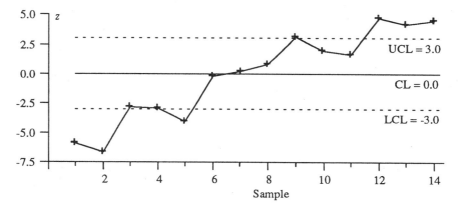

The preceding control chart reveals a definite upward trend, which may be indicating process degradation. Points associated with samples 1, 2, and 5 are below LCL; those associated with samples 9, 12, 13, and 14 are above UCL. A comparison of process conditions at the times samples 1, 2, and 5 were obtained with those at the times associated with samples 9, 12, 13, and 14 may provide insight into the cause of the degradation.

8.61 $\bar{u} = \Sigma c_i / \Sigma a_i = 157/6330 \approx 0.025$. For the ith sample, the standardized value of u_i is $z_i = (u_i - 0.025)\sqrt{a_i}/\sqrt{0.025}$. Using this formula with the data in Problem 8.60 gives the following table.

Sample	1	2	3	4	5	6	7	8
c	1	4	12	6	3	6	1	2
a	100	200	300	300	150	200	100	200
u	0.01	0.02	0.04	0.02	0.02	0.03	0.01	0.01
z	-0.95	-0.45	1.64	-0.55	-0.39	0.45	-0.95	-1.34

107

Sample	9	10	11	12	13	14	15	16
c	12	9	9	2	6	8	12	6
a	300	300	180	200	200	200	300	100
u	0.04	0.03	0.05	0.01	0.03	0.04	0.04	0.06
z	1.64	0.55	2.12	-1.34	0.45	1.34	1.64	2.21

Sample	17	18	19	20	21	22	23	24
c	0	3	2	9	7	4	3	3
a	250	100	200	300	350	200	150	300
u	0.00	0.03	0.01	0.03	0.02	0.02	0.02	0.01
z	-2.50	0.32	-1.34	0.55	-0.59	-0.45	-0.39	-1.64

Sample	25	26	27	28	29	30
c	3	1	6	10	1	6
a	300	100	200	250	100	200
u	0.01	0.01	0.03	0.04	0.01	0.03
z	-1.64	-0.95	0.45	1.50	-0.95	0.45

The following standardized chart contains no signal of the presence of special cause variation. The average count per unit area of opportunity appears to be in control.

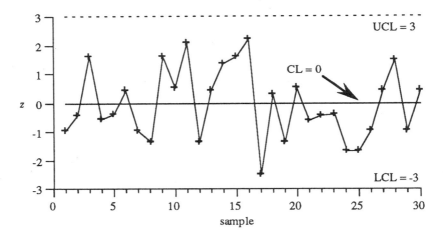

8.63　$\bar{u} = \Sigma c_i / \Sigma a_i = 394 / 197 = 2.0$. For the ith sample, the standardized value of u_i is $z_i = (u_i - 2.0)\sqrt{a_i}/\sqrt{2.0}$. Using this formula with the data in Problem 8.62 gives the following table. From the accompanying control chart, the results of samples 11 and 12 indicate the presence of special cause variation.

Sample	1	2	3	4	5	6	7	8
c	15	10	21	17	6	8	19	9
a	8	8	12	8	8	4	7	8
u	1.9	1.2	1.8	2.1	0.8	2.0	2.7	1.1
z	-0.2	-1.6	-0.5	0.2	-2.4	0.0	1.3	-1.8

Sample	9	10	11	12	13	14	15	16
c	14	8	31	51	3	18	9	23
a	8	8	8	16	4	9	7	8
u	1.8	1.0	3.9	3.2	0.8	2.0	1.3	2.9
z	-0.4	-2.0	3.8	3.4	-1.7	0.0	-1.3	1.8

Sample	17	18	19	20	21	22	23	24
c	14	12	15	13	23	22	16	17
a	4	8	8	8	12	9	9	8
u	3.5	1.5	1.9	1.6	1.9	2.4	1.8	2.1
z	2.1	-1.0	-0.2	-0.8	-0.2	0.8	-0.4	0.2

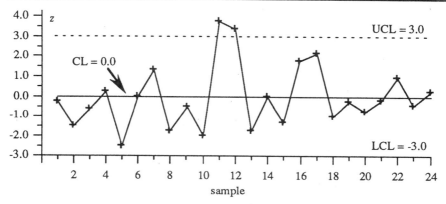

CHAPTER 9
MEASUREMENT SYSTEM EVALUATION

Section 9.1

9.1 **(a)**

Operator	Range of Measurements for Part:									
	1	2	3	4	5	6	7	8	9	10
1	0.01	0.02	0.00	0.02	0.01	0.02	0.03	0.00	0.01	0.00
2	0.02	0.02	0.00	0.00	0.01	0.02	0.01	0.02	0.00	0.01
3	0.01	0.03	0.01	0.02	0.00	0.01	0.03	0.00	0.01	0.00

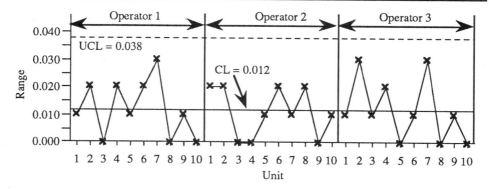

R chart: $CL = \overline{R} = (0.01 + 0.02 + \ldots + 0.00)/30 = 0.35/30 \approx 0.012$, $D_3 = 0$, $D_4 = 3.267$, $LCL = D_3\overline{R} = 0.000$, and $UCL = D_4\overline{R} = (3.267)(0.35/30) \approx 0.038$. All ranges are within the control limits.

(b) $\hat{\sigma}_\varepsilon = \overline{R}/d_2^* \approx 0.012/1.13$

$EV = 5.15\hat{\sigma}_\varepsilon \approx (5.15)(0.012/1.13) \approx 0.05469$

$\%EV_{tolerance} = 100(EV)/(USL - LSL)$

$\qquad \approx 100(0.05469)/(2.835 - 2.565) \approx 20.3$

Section 9.2

9.3 Each of the $J = 3$ operators obtained $n = 2$ measurements for each of $m = 10$ units.

$\overline{y}_1 = (2.71 + 2.70 + 2.72 + 2.70 + \ldots + 2.70 + 2.71 + 2.70 + 2.70)/20$

$\qquad = 54.04/20 = 2.7020$,

$\overline{y}_2 = (2.71 + 2.69 + 2.72 + 2.70 + \ldots + 2.70 + 2.70 + 2.70 + 2.71)/20$

$\qquad = 54.03/20 = 2.7015$,

$\overline{y}_3 = (2.71 + 2.70 + 2.72 + 2.69 + \ldots + 2.70 + 2.71 + 2.70 + 2.70)/20$

$\qquad = 54.04/20 = 2.7020$,

$R_{\overline{y}} = 2.7020 - 2.7015 = 0.0005$, $\hat{\sigma}_\varepsilon \approx 0.012/1.13$ from Problem 9.1, and

$d_2^* = 1.91$ for the 1 sample of 3 means.

Since $\hat{\sigma}_o^2 = \left(R_{\overline{y}}/d_2^*\right)^2 - \left(\hat{\sigma}_\varepsilon^2/mn\right) \approx (0.0005/1.91)^2 - (0.012/1.13)^2/20$ is

negative, let $\hat{\sigma}_o^2 = 0$. This gives $AV = 5.15\hat{\sigma}_o \approx (5.15)(0) = 0$ and

$\%AV_{tolerance} = 0$.

Section 9.3

9.5 $EV \approx 0.05469$ and $AV \approx 0$, from Problems 9.1 and 9.3, respectively.

$R\&R = \sqrt{(AV)^2 + (EV)^2} \approx \sqrt{0^2 + (0.05469)^2} = 0.05469$

$\%R \& R_{tolerance} = 100(R\&R)/(USL - LSL)$

$\qquad \approx 100(0.05469)/(2.835 - 2.565) \approx 20.3$

The 99% interval associated with the repeatability and reproducibility of the measurement system consumes approximately 20.3% of the tolerance spread. The acceptability of this percentage depends upon factors such as cost and application.

110

9.7

Part	Averages for Operator: 1	2	3	Part	Averages for Operator: 1	2	3
1	2.705	2.700	2.705	6	2.700	2.690	2.695
2	2.710	2.710	2.705	7	2.705	2.695	2.705
3	2.700	2.700	2.705	8	2.700	2.710	2.700
4	2.700	2.710	2.700	9	2.705	2.700	2.705
5	2.695	2.695	2.700	10	2.700	2.705	2.700

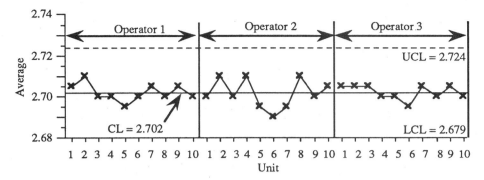

The sums of the 20 observations for operators 1, 2, and 3 are 54.04, 54.03, and 54.04, respectively (Problem 9.3).

$$CL = \bar{\bar{x}} = (54.04 + 54.03 + 54.04)/60 = 162.11/60 \approx 2.702$$

Since $A_2 = 1.880$ for samples of size 2 and $\bar{R} \approx 0.012$ from Problem 9.1,

$$LCL = \bar{\bar{x}} - A_2\bar{R} \approx 2.702 - (1.880)(0.012) = 2.702 - 0.02256 \approx 2.679 \text{ and}$$

$$UCL = \bar{\bar{x}} + A_2\bar{R} \approx 2.702 + (1.880)(0.012) = 2.702 + 0.02256 \approx 2.724.$$

Since all means are between LCL and UCL, the measurement system cannot adequately detect part-to-part variation.

Section 9.6

9.11 We will follow the steps in Table 9.4. Since steps 1 through 7 were completed for us, we begin with step 8. Steps are denoted S8, S9, ...

S8

Operator	Range of Measurements for Module: 1	2	3	4	5	6	7
1	0.003	0.002	0.003	0.000	0.002	0.003	0.002
2	0.000	0.002	0.003	0.002	0.000	0.003	0.005
3	0.002	0.003	0.003	0.003	0.003	0.002	0.003

Operator	Range of Measurements for Module: 8	9	10	11	12	13	14
1	0.000	0.002	0.000	0.002	0.000	0.002	0.000
2	0.002	0.002	0.000	0.000	0.002	0.005	0.000
3	0.002	0.000	0.002	0.003	0.003	0.003	0.000

111

Operator	Range of Measurements for Module:					
	15	16	17	18	19	20
1	0.000	0.000	0.000	0.000	0.003	0.002
2	0.003	0.003	0.000	0.003	0.003	0.005
3	0.000	0.003	0.002	0.000	0.000	0.000

S9   (R chart using all data) CL = $\overline{R}$ = (0.003 + ... + 0.000)/60 = 0.106/60 ≈ 0.0018, D_3 = 0, and D_4 = 2.575. So, LCL = $D_3 \overline{R}$ = 0.0000 and UCL = $D_4 \overline{R}$ = (2.575)(0.106/60) ≈ 0.0045. The ranges of the readings obtained by operator 2 on modules 7, 13, and 20 exceed UCL. All other ranges are between LCL and UCL.

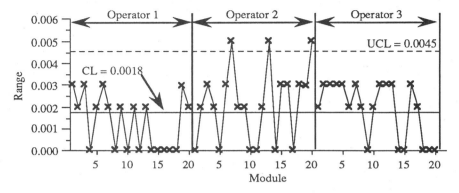

(*R* chart after removing the data associated with operator 2)

Assuming operator 2 needs further training, we will delete the operator 2 data and continue using only that for operators 1 and 3.

The sum of the remaining 40 ranges is 0.063. Letting $\overline{R}$ = 0.063/40 ≈ 0.0016, control limits for the modified set of ranges are LCL = 0.0000, CL = $\overline{R}$ ≈ 0.0016, and UCL = (2.575)(0.063/40) ≈ 0.0041. All ranges are between LCL and UCL.

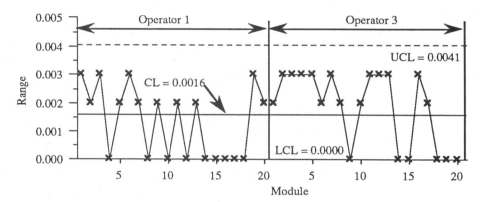

112

S10 For 40 samples of size 3, $d_2^* = 1.69$.

$$\hat{\sigma}_\varepsilon = \overline{R}/d_2^* \approx 0.0016/1.69 \approx 0.0009$$

S11 $EV = 5.15\hat{\sigma}_\varepsilon = (5.15)(0.0016/1.69) \approx 0.004876$

$\%EV_{tolerance} = 100(EV)/(USL - LSL)$
$\approx 100(0.004876)/(0.750 - 0.000) \approx 0.65$

S12 Each of the $J = 2$ operators obtained $n = 3$ measurements for each of $m = 20$ units.

$$\overline{y}_1 = (0.559 + 0.559 + \ldots + 0.584 + 0.584)/60 = 34.423/60 \approx 0.57372$$

$$\overline{y}_3 = (0.556 + 0.558 + \ldots + 0.583 + 0.583)/60 = 34.378/60 \approx 0.57297$$

S13 The range of $\{\overline{y}_1, \overline{y}_3\} = \{0.57372, 0.57297\}$ is

$R_{\overline{y}} = 0.57372 - 0.57297 = 0.00075$.

S14 $\hat{\sigma}_\varepsilon \approx 0.0016/1.69$ from step 10; $d_2^* = 1.41$ for the 1 sample of 2 means

$$\hat{\sigma}_O^2 = \left(R_{\overline{y}}/d_2^*\right)^2 - \left(\hat{\sigma}_\varepsilon^2/mn\right)$$
$$= \left(0.00075/1.41\right)^2 - \left(0.0016/1.69\right)^2\Big/60 \approx 0.00000027$$

S15 $AV = 5.15\,\hat{\sigma}_O \approx 5.15\sqrt{0.00000027} \approx 0.0027$

$\%AV_{tolerance} = (100)(AV)/(USL - LSL)$
$\approx 100(0.0027)/(0.750 - 0.000) \approx 0.36$

S16 $\hat{\sigma}_\varepsilon = 0.0016/1.69$ from step 10, so $\hat{\sigma}_\varepsilon^2 \approx 0.00000090$.

$$\hat{\sigma}_M^2 = \hat{\sigma}_O^2 + \hat{\sigma}_\varepsilon^2 \approx 0.00000027 + 0.00000090 = 0.00000117$$

S17 $R\&R = 5.15\,\hat{\sigma}_M \approx 5.15\sqrt{0.00000117} \approx 0.0056$

$\%R\&R_{tolerance} = 100(R\&R)/(USL - LSL) \approx 100(0.0056)/0.750 \approx 0.75$

S18 A table of the part averages for operators 1 and 3 follows.

Part	Operator 1	Operator 3	Part	Operator 1	Operator 3	Part	Operator 1	Operator 3
1	0.558	0.557	8	0.569	0.567	15	0.606	0.606
2	0.540	0.539	9	0.595	0.593	16	0.589	0.590
3	0.607	0.605	10	0.559	0.557	17	0.579	0.577
4	0.564	0.564	11	0.565	0.564	18	0.576	0.576
5	0.565	0.564	12	0.569	0.569	19	0.562	0.561
6	0.587	0.587	13	0.550	0.549	20	0.585	0.583
7	0.575	0.575	14	0.576	0.576			

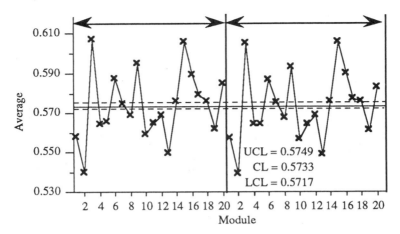

The sums of the 60 observations for operators 1 and 3 are 34.423 and 34.378, respectively (step 12). Thus, $\overline{\overline{x}} = (34.423 + 34.378)/120 \approx 0.5733$. Since $A_2 = 1.023$ for samples of size 3 and $\overline{R} = 0.063/40$ from step 9, LCL $= \overline{\overline{x}} - A_2\overline{R} \approx 0.5733 - (1.023)(0.063/40) \approx 0.5717$ and UCL $= \overline{\overline{x}} + A_2\overline{R} \approx 0.5733 + (1.023)(0.063/40) \approx 0.5749$ are the control limits for $\overline{x}$. Only the mean of the sample obtained by operator 1 for module 7 is between the control limits. [Before rounding, that mean is $1.724/3 = 0.57466...$.] The measurement system can detect part-to-part variability.

S19   Totals and averages (to the nearest ten thousandth) for the 6 readings (3 by each of operators 1 and 3) on each of the 20 modules are summarized in the following table.

Module	Total	Average	Module	Total	Average
1	3.346	0.5577	11	3.386	0.5643
2	3.236	0.5393	12	3.414	0.5690
3	3.636	0.6060	13	3.296	0.5493
4	3.384	0.5640	14	3.456	0.5760
5	3.388	0.5647	15	3.636	0.6060
6	3.521	0.5868	16	3.537	0.5895
7	3.449	0.5748	17	3.469	0.5782
8	3.409	0.5682	18	3.456	0.5760
9	3.563	0.5938	19	3.369	0.5615
10	3.347	0.5578	20	3.503	0.5838

$R_{\text{part means}} \approx 0.6060 - 0.5393 = 0.0667 \approx 0.067$ and $d_2^* \approx 3.805$ for 1 sample of size 20. Thus, $\hat{\sigma}_P^2 = \left(R_{\text{part means}}\middle/d_2^*\right)^2 \approx (0.067/3.805)^2$ and $\hat{\sigma}_P \approx 0.067/3.805 \approx 0.0176$.

114

$PV = 5.15\hat{\sigma}_p \approx (5.15)(0.0176) \approx 0.0906$

$R\&R \approx 0.0056$ from step 17.

$TV = \sqrt{(PV)^2 + (R\&R)^2} \approx \sqrt{(0.0906)^2 + (0.0056)^2} \approx 0.0908$

$\%R\&R_{process} = 100(R\&R)/TV \approx 100(0.0056)/0.0908 \approx 6.17.$

The repeatability and reproducibility of the measurement system accounts for about 6.17% of the total variabilty.

S21   Both $\%R\&R_{tolerance} \approx 0.75$ and $\%R\&R_{process} \approx 6.17$ are acceptable.

CHAPTER 10
PROCESS CAPABILITY STUDIES

Section 10.1

10.1   The following box plot reveals that 0.491 inch is a strong outlier. A quick scan of the data, histogram, or normal quantile plot will reveal the same information.

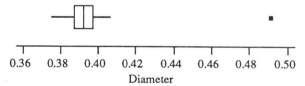

The following outputs are obtained after deleting 0.491 (an obvious outlier) and using JMP with the remaining 59 values. Approximately 32% of the observed values are below LSL = 0.390. No value is above USL = 0.410.

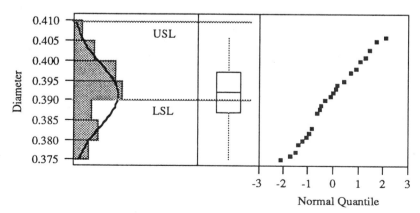

Moments	
Mean	0.39137
Std Dev	0.00783
Std Err Mean	0.00102
upper 95% Mean	0.39341
lower 95% Mean	0.38933
N	59.00000

Test for Normality	
Shapiro-Wilk W Test	
W	Prob<W
0.957345	0.0805

Specification	Value	Percent	Actual	Normal
Lower Spec Limit	0.39	%Below LSL	32.203	43.039
Upper Spec Limit	0.41	%Above USL	0.000	0.866

The p-value of the Shapiro-Wilk W test for normality is 0.0805, the normal quantile plot is only slightly S-shaped, the box plot reveals no outliers, and the histogram is reasonably bell-shaped. A normality assumption may be valid.

Continuing with the JMP outputs, $\bar{x} \approx 0.3914$ and $s \approx 0.0078$. *If* the population of diameters produced during the study has an approximate normal distribution with mean $\mu = 0.3914$ and standard deviation $\sigma = 0.0078$, approximately 43.1% of the population values are below LSL = 0.390. [Using $\bar{x}$ and s to more decimal places, JMP reports 43.039%.] Approximately 0.9% are above USL = 0.410. At the time of the study, approximately 44% of all shafts produced did not meet specifications.

10.3 If the range is not in control, we have no valid estimate of a single population's standard deviation.

Section 10.2

10.5

Sample	1	2	3	4	5	6	7	8	9	10
Range	0.18	0.18	0.54	0.18	0.36	0.54	0.36	0.54	0.18	0.36
Sample	11	12	13	14	15	16	17	18	19	20
Range	0.54	0.54	0.36	0.36	0.72	0.36	0.54	0.36	0.54	0.90

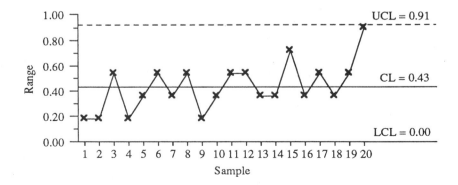

R chart: $\overline{R} = (0.18 + 0.18 + \ldots + 0.54 + 0.90)/20 = 8.64/20 = 0.432$ for the set of ranges in the preceding table. Thus, CL = $\overline{R}$ = 0.432 ≈ 0.43. For samples of size 5, $D_3 = 0$ and $D_4 = 2.115$. Control limits for R are LCL = $D_3\overline{R} = 0.00$ and UCL = $D_4\overline{R} \approx (2.115)(0.432) \approx 0.91$. The 20 ranges are between these control limits.

For samples of size 5, $d_2 = 2.326$. Thus, $\hat{\sigma}_c = \overline{R}/d_2 = 0.432/2.326 \approx 0.186$. The process capability is $6\hat{\sigma}_c \approx 6(0.186) = 1.12$. Since USL - LSL = 5.76 - 4.50 = 1.26 is greater than $6\hat{\sigma}_c$, the process may be capable of meeting specifications.

10.9 Control limits for the range are based on a standard of $\sigma = \hat{\sigma}_c = 0.432/2.326 \approx 0.186$. For samples of size 5, $D_1 = 0$, $d_2 = 2.326$, and $D_2 = 4.918$. Thus, LCL = $D_1\sigma = 0.00$, CL = $d_2\sigma = (2.326)(0.432/2.326) \approx 0.43$ and UCL = $D_2\sigma = (4.918)(0.432/2.326) \approx 0.91$ are control limits for R.

If the process can be centered at x_{nom} = 5.13 degrees, control limits for $\overline{x}$ are based on standards of μ = 5.13 and σ = 0.186. For samples of size 5, A = 1.342. This gives LCL = $\mu - A\sigma$ = 5.13 - (1.342)(0.186) ≈ 4.88, CL = μ = 5.13, and UCL = $\mu + A\sigma$ ≈ 5.38.

Suppose the process cannot (or should not) easily be changed.

Sample	1	2	3	4	5	6	7
Total	26.64	26.46	26.46	26.64	26.82	26.28	26.82
$\overline{x}$	5.328	5.292	5.292	5.328	5.364	5.256	5.364

Sample	8	9	10	11	12	13	14
Total	26.10	26.64	26.10	25.92	27.00	26.46	26.82
$\overline{x}$	5.220	5.328	5.220	5.184	5.400	5.292	5.364

Sample	15	16	17	18	19	20
Total	26.10	26.10	27.18	26.64	25.74	26.28
$\overline{x}$	5.220	5.220	5.436	5.328	5.148	5.256

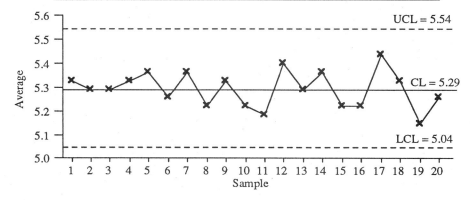

Using the totals in the preceding table, $\overline{\overline{x}} = (26.64 + \ldots + 26.28)/100 =$ 529.20/100 = 5.292. Using $\sigma = \hat{\sigma}_c = 0.432/2.326 \approx 0.186$ as a standard and $A = 1.342$, control limits for $\overline{x}$ are LCL $= \overline{\overline{x}} - A\sigma \approx 5.292 - (1.342)(0.186) \approx$ 5.04 and UCL $= \overline{\overline{x}} + A\sigma \approx 5.292 + (1.342)(0.186) \approx 5.54$. The preceding $\overline{x}$ chart contains no signal of the presence of assignable causes, so these limits can be used to monitor the sample average.

Section 10.4

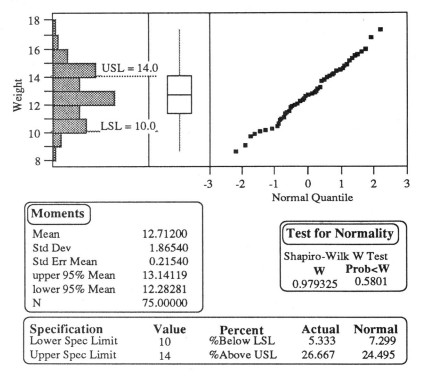

Consider the preceding JMP outputs. The histogram, box plot, normal quantile plot, and Shapiro-Wilk W test (p-value = 0.5801) support the assumption that the population of weights produced during the study is normally distributed.

Using the specification summary, 4 of the 75 observed values (approximately 5.3%) are less than LSL and 20 of those values (approximately 26.7%) are greater than USL. If the sampled population is normally distributed with mean $\mu \approx \overline{x} = 12.712$ and standard deviation $\sigma = s \approx 1.865$, the percentages of weights that do not meet specifications are 7.3% below LSL and 24.5% above USL. The process was not meeting specifications at the time of the study.

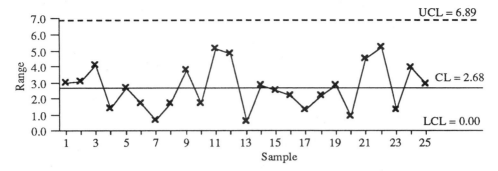

Consider the preceding JMP range chart. All points are within the control limits. No obvious trends or patterns are present. It appears that the range was in control at the time of the study.

For samples of size 3, $d_2 = 1.693$. Since CL = $\overline{R} \approx 2.7$, $\hat{\sigma}_c = \overline{R}/d_2 \approx$ 2.7/1.693 $\approx$ 1.59. The process capability is $6\hat{\sigma}_c \approx 6(1.59) = 9.54$. Since the allowable spread is only 4 units, the process is not capable of meeting specifications.

Suppose the process can be centered at $\mu = 12.00$ and the standard deviation can be maintained at $\sigma = 1.59$. *If* the population of weights is normally distributed at any point in time, then $P(X \leq 10) + P(X \geq 14) \approx$ $P(Z \leq -1.26) + P(Z \geq 1.26) \approx 0.20766$. Thus, about 21% of the weights will fail to meet specifications. Since $D_1 = 0$, $d_2 = 1.693$, $D_2 = 4.358$, and $A = 1.732$ for samples of size 3, the process can then be monitored using the following control limits:

Range chart: LCL = 0.000, CL $\approx$ (1.693)(1.59) $\approx$ 2.69, and
UCL = (4.358)(1.59) $\approx$ 6.93.

Averages chart: LCL = 12.00 - (1.732)(1.59) $\approx$ 12.00 - 2.75 = 9.25, CL = 12.00, and UCL $\approx$ 12.00 + 2.75 = 14.75.

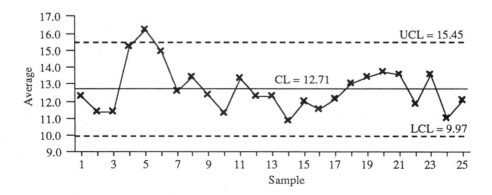

119

Using JMP, the preceding $\bar{x}$ chart of the original data is obtained. The mean of sample 5 exceeds UCL, so $\bar{x}$ was not in control during the study.

Removing sample 5, the following JMP chart of the averages results. The 24 samples seem to have been obtained from a population with mean $\mu \approx 12.6$. The proportion of out-of-specification weights is greater when $\mu = 12.6$ than that when $\mu = 12.0$. Use of control charts to monitor a process with $\mu = 12.6$ and $\sigma = 1.59$ is unlikely.

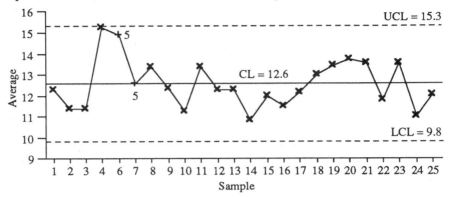

[Note: JMP includes some tests for process centering that are not discussed in the text. One of those, indicated by the numeral 5 on an $\bar{x}$ chart, occurs when at least 2 of 3 consecutive points on the same side of the center line are over 2 standard deviations from the mean. Such circumstances often indicate a shift in the process centering.

Two such signals are indicated on the preceding $\bar{x}$ chart. The first indicates that at least 2 of the means for samples 3, 4, and 6 are more than 2 standard deviations from the mean. The second indicates that at least 2 of the means for samples 4, 6, and 7 are more than 2 standard deviations from the mean.]

10.17 Using JMP, the following graphics and summary statistics were obtained for the set of 125 observed values. As with the relative frequency histogram obtained when solving Problem 2.9, it appears that the population (*at the time of the study*) is skewed to the right. The histogram, box plot, normal quantile plot, and Shapiro-Wilk W test (*p*-value = 0.0006) support an assumption of nonnormality. The indicated outliers appear to be consistent with a skewed distribution.

All observed values are between 0.0 and 10.0 mils. However, the study was conducted to determine whether the upper specification could be changed from 10 mils to 6 mils. From the following specification summary, 5 of the 125 observed values (or 4%) of the exceed 6.0 mils. At the time of the study, the process was not meeting that proposed upper specification limit. [Note: In Problem 2.9, an ogive was constructed and used to estimate that 7% of the sampled population exceeded 6.0 mils.]

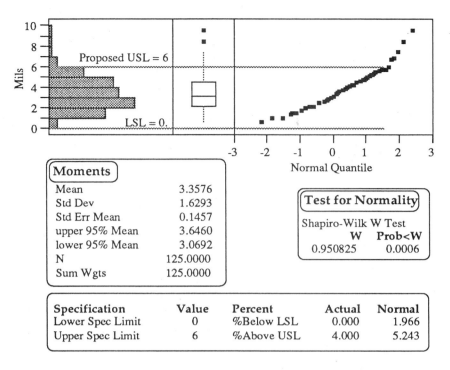

Moments	
Mean	3.3576
Std Dev	1.6293
Std Err Mean	0.1457
upper 95% Mean	3.6460
lower 95% Mean	3.0692
N	125.0000
Sum Wgts	125.0000

Test for Normality

Shapiro-Wilk W Test

W	Prob<W
0.950825	0.0006

Specification	Value	Percent	Actual	Normal
Lower Spec Limit	0	%Below LSL	0.000	1.966
Upper Spec Limit	6	%Above USL	4.000	5.243

From the moments portion of the preceding summary, the sample mean and standard deviation of the set of 125 values are 3.3576 mils and 1.6293 mils, respectively. The histogram indicates that the sampled population is unimodal. Assuming $\mu \approx 3.3576$ and $\sigma \approx 1.6293$ and applying the Camp-Meidel inequality, at least 95.06% of the population values are in the interval $\mu \pm 3\sigma \approx [0, 8.25]$.

The range chart in the solution of Problem 8.19 indicates that the range of sample 11 is outside the control limits. When that sample is removed and control limits are recalculated, the following JMP chart results. All remaining ranges are within the new control limits.

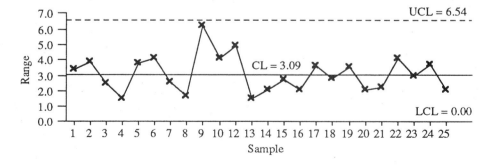

For samples of size 5, $d_2 = 2.326$. Since CL = $\overline{R}$ = 3.09, $\hat{\sigma}_c = \overline{R}/d_2 =$ 3.09/2.326 ≈ 1.33 mils. The process capability is $6\hat{\sigma}_c \approx 6(1.33) = 7.98$ mils. Since the allowable spread is 10 mils, the process may be capable of meeting specifications of 0 to 10 mils.

The study was conducted to determine whether the upper specification could be changed from 10 mils to 6 mils. Since the process capability is 7.98 mils, the process is not capable of meeting modified specifications of 0 to 6 mils.

To study the process centering, we first look at an $\overline{x}$ chart based on the 24 samples remaining after sample 11 is removed. In the following JMP chart for those samples, the mean of sample 9 exceeds UCL.

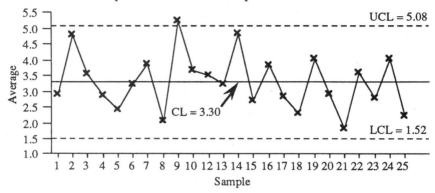

Removing sample 9, the following $\overline{x}$ and R charts for the remaining 23 samples result. No out-of-control condition is indicated on either chart. If the samples are representative of the process, the control limits for these charts can be extended to monitor the process in the near future.

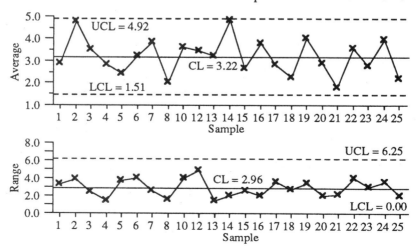

122

10.23 USL = 5.76, LSL = 4.50, $\hat{\sigma}_c \approx 0.186$ from Problem 10.5; and $\bar{\bar{x}} \approx 5.29$ from Problem 10.9.

$$\hat{C}_p = (USL - LSL)/(6\hat{\sigma}_c) \approx (5.76 - 4.50)/[6(0.186)] \approx 1.13$$
$$\hat{C}_{pL} = (\bar{\bar{x}} - LSL)/(3\hat{\sigma}_c) \approx (5.29 - 4.50)/[3(0.186)] \approx 1.42$$
$$\hat{C}_{pU} = (USL - \bar{\bar{x}})/(3\hat{\sigma}_c) \approx (5.76 - 5.29)/[3(0.186)] \approx 0.84$$
$$\hat{C}_{pk} = \min\{\hat{C}_{pL}, \hat{C}_{pU}\} = \min\{1.42, 0.84\} = 0.84$$

Since $\hat{C}_p > 1$, the process is capable of meeting specifications.

$\hat{C}_{pU} < \hat{C}_{pL}$, so the process is centered to the right of $x_{nom} = 5.13$.

Since $\hat{C}_{pU} < 1$, the process may be producing products that do not meet the upper specification limit.

10.27 USL = 14.0, LSL = 10.0, $\hat{\sigma}_c \approx 1.59$ (Problem 10.13), and $\bar{\bar{x}} \approx 12.6$ (after deleting sample 5).

$$\hat{C}_p = (USL - LSL)/(6\hat{\sigma}_c) \approx (14.0 - 10.0)/[6(1.6)] \approx 0.42$$
$$\hat{C}_{pL} = (\bar{\bar{x}} - LSL)/(3\hat{\sigma}_c) \approx (12.6 - 10.0)/[3(1.6)] \approx 0.54$$
$$\hat{C}_{pU} = (USL - \bar{\bar{x}})/(3\hat{\sigma}_c) \approx (14.0 - 12.6)/[3(1.6)] \approx 0.29$$
$$\hat{C}_{pk} = \min\{\hat{C}_{pL}, \hat{C}_{pU}\} = \min\{0.54, 0.29\} = 0.29$$

Since $\hat{C}_p < 1$, the process is not capable of meeting specifications.

$\hat{C}_{pU} < \hat{C}_{pL}$, so the process is centered to the right of $x_{nom} = 12.0$.

Since $\hat{C}_{pU} < 1$, the process may be producing products that do not meet the upper specification limit.

$\hat{C}_{pL} < 1$, so the process may be producing products that do not meet the lower specification limit.

10.35 **(a)** Since the initial setup was in error (Example 8.4), the values in sample 1 may not be representative of the process. The following JMP outputs and remarks apply to the data set consisting of samples 2 through 18.

The histogram, box plot, and normal quantile plot do not seem to give strong evidence against an assumption of normality. However, the p-value of the Shapiro-Wilk W test (p-value =

0.0000) is so small that an assumption of nonnormality seems warranted. We will assume that the population of flatness readings associated with the housings produced at the time of the study has a nonnormal distribution.

Using the specifications summary, 28 of the 85 values (about 32.9%) are greater than USL. The stated capability index (CPU = 0.190) was obtained using s instead of $\hat{\sigma}_c$ in the text formula for $\hat{C}_{pU}$. That is, $\text{CPU} = \left(\text{USL} - \overline{\overline{x}}\right)/(3s)$ with $\overline{\overline{x}} = 0.23292$ and $s = 0.02995$. The small value of CPU indicates that the sample mean is very close to the upper specification limit.

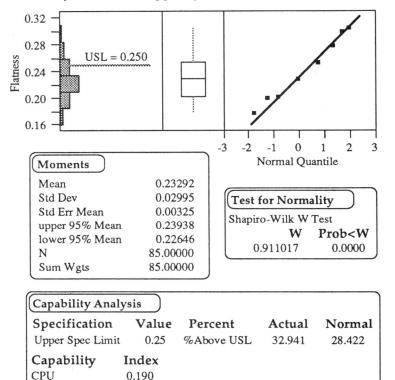

Moments	
Mean	0.23292
Std Dev	0.02995
Std Err Mean	0.00325
upper 95% Mean	0.23938
lower 95% Mean	0.22646
N	85.00000
Sum Wgts	85.00000

Test for Normality		
Shapiro-Wilk W Test		
	W	Prob<W
	0.911017	0.0000

Capability Analysis				
Specification	Value	Percent	Actual	Normal
Upper Spec Limit	0.25	%Above USL	32.941	28.422
Capability	Index			
CPU	0.190			

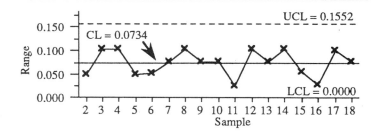

124

Consider the preceding JMP range chart. No unusual trend or run is present and no point is outside the control limits. For samples of size 5, $d_2 = 2.326$. Since CL $= \overline{R} \approx 0.073$, $\hat{\sigma}_c = \overline{R}/d_2 = 0.073/2.326 \approx 0.0314$. The process capability is $6\hat{\sigma}_c \approx 6(0.0314) \approx 0.1884$.

The following $\overline{x}$ chart seems to indicate that the process centering was in control with $\mu \approx$ CL $= \overline{\overline{x}} \approx 0.2329$. If the process standard deviation can be maintained at $\sigma \approx \hat{\sigma}_c \approx 0.0314$ and the process mean can be shifted from $\mu \approx 0.2329$ to USL $- 4\sigma \approx$ $0.2500 - 4(0.0314) = 0.1244$, the proportion of housings with out-of-specification flatness readings will be less than 0.0278. This estimate is obtained using the Camp-Meidel Inequality stated in Section 2.4.4.

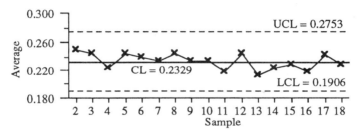

(b) From (a), $\hat{\sigma}_c \approx 0.0314$ and $\overline{\overline{x}} \approx 0.2329$. Since USL $= 0.250$,

$$\hat{C}_{pU} = \left(\text{USL} - \overline{\overline{x}}\right)\big/\left(3\hat{\sigma}_c\right) \approx (0.2500 - 0.2329)/[3(0.0314)] \approx 0.18.$$

This small value indicates that the data are centered very close to the upper specification limit and that the process may be producing housings with flatness readings greater than USL.

10.39 Using the 59 values remaining after the obvious outlier is removed, $s \approx 0.00783$. Thus,

CP $=$ (USL - LSL)$/(6s) \approx (0.410 - 0.390)/[6(0.00783)] \approx 0.4257$.

The 10th percentile of a chi-square distribution with 58 degrees of freedom is 44.696. [**Note:** Using linear interpolation and Appendix F, 44.699 is obtained.] Hence,

$$L = (\text{CP})\sqrt{\chi^2_{0.10}\big/(n-1)} \approx 0.4257\sqrt{44.696/58} \approx 0.374$$

is a 90% lower confidence limit for C_p [Equation (10.10)].

CHAPTER 11

AN INTRODUCTION TO REGRESSION ANALYSIS

Section 11.1

11.1 **(a)** Since $E[Y \mid x] = 4 + 7.5x$, the average Schopper-Riegler reading at $x = 7$ is $4 + (7.5)(7) = 56.5$ degrees.

(b) The slope of the line of the means is $\beta_1 = 7.5$. Thus, a one hour increase in the beating time produces a 7.5 degree increase in the average Schopper-Riegler reading.

(c) From (a), the mean of Y at $x = 7$ is 56.5 degrees. So, $Y \sim N(56.5, 28.8)$ at $x = 7$. Thus,

$$
\begin{aligned}
P(45.5 \le Y \le 59.5) &= P(Y \le 59.5) - P(Y \le 45.5) \\
&= P[Z \le (59.5 - 56.5)/\sqrt{28.8}] \\
&\quad - P[Z \le (45.5 - 56.5)/\sqrt{28.8}] \\
&\approx \Phi(0.56) - \Phi(-2.05) \\
&= 0.71226 - 0.02018 = 0.69208
\end{aligned}
$$

(d) At $x = 4$, the mean of Y is $4 + (7.5)(4) = 34.0$. So, $Y \sim N(34.0, 28.8)$ at $x = 4$. Thus,

$$
\begin{aligned}
P(23 \le Y \le 45) &= P(Y \le 45) - P(Y \le 23) \\
&= P[Z \le (45 - 34.0)/\sqrt{28.8}] \\
&\quad - P[Z \le (23 - 34.0)/\sqrt{28.8}] \\
&\approx \Phi(2.05) - \Phi(-2.05) \\
&= 0.97982 - 0.02018 = 0.95964
\end{aligned}
$$

11.3 **(a)** Since $E[Y \mid x] = -2.16 + 10.67x$, $-2.16 + (10.67)(3.5) = 35.185$ is the average value of Y at $x = 3.5$. So, $Y \sim N(35.185, 0.01)$. Thus,

$$
\begin{aligned}
P(Y > 35.335) &= P[Z > (35.335 - 35.185)/\sqrt{0.01}] \\
&= P(Z > 1.50) = \Phi(-1.50) = 0.06681.
\end{aligned}
$$

(b) Let W denote the number of y values that exceed 35.335. Since $P(Y > 35.335) = 0.06681$ at $x = 3.5$, $W \sim b(4, 0.06681)$. Thus, $P(W = 1) = C(4, 1) \times (0.06681)(0.93319)^3 \approx 0.21718$.

Section 11.2

11.5 **(a)** $b_0 \approx 1.13$; $b_1 \approx 14.49$

(b) The points in the following scatter plot are near the least squares line. There seems to be a strong, linear relationship between the tensile force and the elongation.

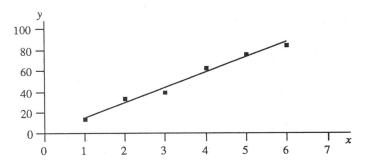

(c) From (a), $\hat{y} = 1.13 + 14.49x$. At $x = 4$, the average elongation is approximately $1.13 + (14.49)(4) = 59.09$.

(d) Using $\hat{y} = 1.13 + 14.49x$, the six $(\hat{y}, y)$ pairs are (15.62, 14), (30.10, 33), (44.59, 40), (59.08, 63), (73.56, 76), and (88.05, 85).

(e) A plot of the six $(\hat{y}, y)$ pairs follows. The points lie very near the 45-degree line through the origin.

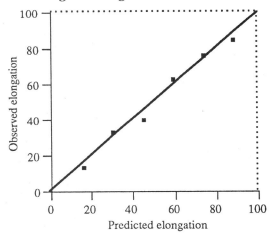

(f) Since the points in the graphic in (e) lie very near the 45-degree line, the difference between an observed value and the corresponding predicted value (i.e., the residual) tends to be small. Thus, the least squares line provides "good" predictions.

11.7 **(a)** $SSX = \sum x_i^2 - \left(\sum x_i\right)^2 / n = 121 - 20^2/10 = 121 - 40 = 81$

$SPXY = \sum x_i y_i - (\sum x_i)(\sum y_i)/n = (-82) - (20)(40)/10 = (-82) - 80 = -162$

$b_1 = (SPXY)/(SSX) = (-162)/81 = -2$

$b_0 = \bar{y} - b_1 \bar{x} = (\sum y_i)/n - b_1(\sum x_i)/n = (40/10) - (-2)(20/10) = 4 - (-4) = 8$

$\hat{y} = b_0 + b_1 x = 8 - 2x$

127

(b) $SSY = \sum y_i^2 - \left(\sum y_i\right)^2 / n = 516 - 40^2/10 = 516 - 160 = 356$. From (a), $SSX = 81$ and $SPXY = -162$. Thus, $r = (SPXY)/\sqrt{(SSX)(SSY)} = (-162)/\sqrt{(81)(356)} = -9/\sqrt{89} \approx -0.95$. Since $r^2 = 81/89 \approx 0.91$, approximately 91% of the variability in the y values is associated with the simple linear model.

11.9 **(a)** $\hat{y} \approx -0.16289 + 0.00046x$

(b) A scatter plot with the least squares line included follows It appears that a different model may be needed. Some curvature seems to be indicated.

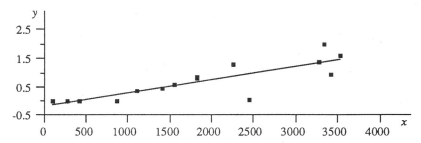

Section 11.3

11.13 **(a)** $s_Y^2 \approx 0.464596$, $r^2 \approx 0.979186$, $n = 10$, and $SSY = (n-1)s_Y^2 \approx 9(0.464596) = 4.181364$.

$SSM = (SSY)r^2 \approx (4.181364)(0.979186) \approx 4.094333$
$SSR = SSY - SSM \approx 4.181364 - 4.094333 = 0.087031$

(b) $MSR = (SSR)/(n-2) \approx 0.087031/8 \approx 0.010879$

(c) When the model assumptions are valid, the calculated value of MSR is an unbiased estimate of the common variance.

(d) $\hat{y} \approx -0.027 + 1.084x$ from Problem 2.55. When $x = 2$, the average time to complete the job is approximately $-0.027 + (1.084)(2) = 2.141 \approx 2.14$ hours.

(e) $\hat{y} \approx -0.027 + 1.084x$ from Problem 2.55. Thus, the next job for which the estimated time is 1.50 hours will take approximately $-0.027 + (1.084)(1.50) = 1.599 \approx 1.60$ hours to complete.

11.15 **(a)** From Problem 11.7, $SSY = 356$ and $r^2 = 81/89$. Thus, $SSM = (SSY)r^2 = (356)(81/89) = 324$ and $SSR = SSY - SSM = 356 - 324 = 32$.

(b) Since $n = 10$ pairs were used, $MSR = (SSR)/(n-2) = 32/8 = 4$. When the model assumptions are valid, 4 is an unbiased estimate of the common variance.

11.17 **(a)** $SSX = (n-1)s_X^2 = 5(3.5) = {}^{35}\!/_2$

$SPXY = \Sigma x_i y_i - (\Sigma x_i \Sigma y_i)/n = 1342 - (21)(311)/6 = {}^{507}\!/_2$

$b_1 = (SPXY)/(SSX) = {}^{507}\!/_{35} \approx 14.49$

$b_0 = \bar{y} - b_1 \bar{x} = \left({}^{311}\!/_6\right) - \left({}^{507}\!/_{35}\right)\left({}^{21}\!/_6\right) = {}^{17}\!/_{15} \approx 1.13$

Using $\hat{y} = \left({}^{17}\!/_{15}\right) + \left({}^{507}\!/_{35}\right)x = \left({}^1\!/_{105}\right)(119 + 1521x)$, we obtain the following table.

i	x_i	y_i	$\hat{y}_i$	$e_i = y_i - \hat{y}_i$	e_i^2
1	1	14	${}^{1640}\!/_{105}$	$-{}^{170}\!/_{105}$	${}^{28,900}\!/_{11,025}$
2	2	33	${}^{3161}\!/_{105}$	${}^{304}\!/_{105}$	${}^{92,416}\!/_{11,025}$
3	3	40	${}^{4682}\!/_{105}$	$-{}^{482}\!/_{105}$	${}^{232,324}\!/_{11,025}$
4	4	63	${}^{6203}\!/_{105}$	${}^{412}\!/_{105}$	${}^{169,744}\!/_{11,025}$
5	5	76	${}^{7724}\!/_{105}$	${}^{256}\!/_{105}$	${}^{65,536}\!/_{11,025}$
6	6	85	${}^{9245}\!/_{105}$	$-{}^{320}\!/_{105}$	${}^{102,400}\!/_{11,025}$
Total	21	311		0	${}^{691,320}\!/_{11,025}$

(b) $s_Y^2 = 746\frac{29}{30} = {}^{22,409}\!/_{30}$

$SSY = (n-1)s_Y^2 \approx 5({}^{22,409}\!/_{30}) = {}^{22,409}\!/_6 \approx 3734.8333$

From the table in (a), $SSR = {}^{691,320}\!/_{11,025} = {}^{6584}\!/_{105} \approx 62.7048$.

$SSM = SSY - SSR = ({}^{22,409}\!/_6) - ({}^{6584}\!/_{105}) = {}^{257,049}\!/_{70} \approx 3672.1286$

(c) $MSR = (SSR)/(n-2) = ({}^{6584}\!/_{105}) \div 4 = 1646/105 \approx 15.6762$

(d) When the model assumptions are valid, the calculated value of MSR is an unbiased estimate of the common variance.

(e) $r^2 \approx 0.9832$. Approximately 98.32% of the variability in elongation is associated with the least squares line.

11.19 **(a)** $b_0 \approx -2.209413777 \approx -2.21$; $b_1 \approx 1.307944094 \approx 1.31$; $\hat{y} \approx -2.21 + 1.31x$

(b) The average pressure is approximately $-2.21 + (1.31)(20) = 23.99$ mm when the temperature is 20 degrees. [**Note:** If this is obtained directly from a calculator with linear regression capabilities, the result is approximately 23.95 mm.]

(c)

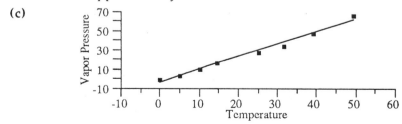

129

The least squares line is included on the preceding scatter plot. The simple linear regression model appears to provide a good fit.

(d) $SSY = (n-1)s_Y^2 \approx 7(530.76125000) = 3715.32875000 \approx 3715.33$

 $r^2 \approx 0.98680974$

 $SSM = (SSY)r^2$

 $\approx (3715.32875000)(0.98680974)$

 $\approx 3666.32259780 \approx 3666.32$

 $SSR = SSY - SSM \approx 3715.32875000 - 3666.32259780$

 $= 49.00615220 \approx 49.01$

(e) $MSR = (SSR)/(n-2) \approx 49.00615220/6 \approx 8.16769203 \approx 8.17$

(f) When the model assumptions are valid, the calculated value of MSR is an unbiased estimate of the common variance.

(g) $r^2 \approx 0.9868$; About 98.68% of the variability in vapor pressure is associated with the least squares line.

Section 11.4

11.21 (a) $H_0: \beta_1 = 0; H_a: \beta_1 \neq 0$

Test statistic: $T = B_1\sqrt{SSX}/\sqrt{MSR}$; $\nu = 4$

Decision rule: Reject H_0 if the p-value is less than α.
(We will use $\alpha = 0.05$.)

Analysis: From Problem 11.12, $b_1 = 116/75$, $SSX = 75/2$, and $MSR = 211/150$. Since $t = [(116/75)\sqrt{75/2}]/\sqrt{211/150} \approx 7.99$, the p-value of the test is $2P(T \geq 7.99) \approx 2(0.0007) = 0.0014$. We reject H_0 and conclude that $\beta_1 \neq 0$.

(b) From Problem 11.12, $b_1 = 116/75$, $SSX = 75/2$, and $MSR = 211/150$. The 97.5th percentile of a t distribution with 4 degrees of freedom is 2.776. Thus,

$b_1 \pm (2.776)\sqrt{MSR}/\sqrt{SSX} = (116/75) \pm (2.776)\sqrt{(211/150)}/\sqrt{(75/2)}$

$\approx 1.5467 \pm 0.5376 \approx [1.01, 2.08]$

is a 95% confidence interval for β_1. With 95% confidence, when x is increased by 1 unit, the increase in the average value of y is at least 1.01 units and at most 2.08 units.

11.23 (a) $H_0: \beta_1 = 0; H_a: \beta_1 \neq 0$

Test statistic: $T = B_1\sqrt{SSX}/\sqrt{MSR}$; $\nu = 4$

Decision rule: Reject H_0 if the p-value is less than α.
(We will use $\alpha = 0.10$.)

Analysis: From Problem 11.17, $SSX = 35/2$, $b_1 = 507/35$, and $SSR = 6584/105$. Since $n = 6$, $MSR = (SSR)/4 = 6584/[(4)(105)] = 1646/105$. So, $t = [(507/35)\sqrt{(35/2)}]/\sqrt{(1646/105)} \approx 15.31$ and the p-value is $2P(T \geq 15.31) \approx 0.0001$. Reject H_0 and conclude that $\beta_1 \neq 0$.

(b) The 95th percentile of a t distribution with 4 degrees of freedom is 2.132. From (a), $b_1 = 507/35$, $SSX = 35/2$, and $MSR = 1646/105$. So,

$$b_1 \pm (2.132)\sqrt{MSR}/\sqrt{SSX} = (507/35) \pm (2.132)\sqrt{(1646/105)}/\sqrt{(35/2)}$$
$$\approx 14.49 \pm 2.02 = [12.47, 16.51]$$ is a 90% confidence interval for β_1.

With 90% confidence, a 1000 pound increase in the force applied increases the average elongation by at least 12.47 thousandths of an inch and at most 16.51 thousandths of an inch.

11.25 (a) $H_0: \beta_1 = 0$; $H_a: \beta_1 \neq 0$; Test statistic: $T = B_1\sqrt{SSX}/\sqrt{MSR}$; $\nu = 9$
Decision rule: Reject H_0 if the p-value is less than α.
(We will use $\alpha = 0.05$.)

Analysis: $SSX = 10 s_X^2 \approx 10(7616.76363636) = 76{,}167.6363636$. From Problem 11.18, $b_1 \approx -0.00128232 \approx -0.0013$ and $MSR \approx 0.000993670/9 \approx 0.00011041 \approx 0.0001$. Since $t \approx (-0.00128232) \frac{\sqrt{76{,}167.6363636}}{\sqrt{0.00011041}} \approx -33.68$, the p-value is $2P(T \leq -33.68) \approx 0.0000$. Reject H_0 and conclude that $\beta_1 \neq 0$.

(b) From (a), $b_1 \approx -0.00128232$, $SSX \approx 76{,}167.6363636$, and $MSR \approx 0.00011041$. Since the 97.5th quantile of a t distribution with 9 degrees of freedom is 2.262,

$$b_1 \pm (2.262)\sqrt{MSR}/\sqrt{SSX} \approx -0.00128232 \pm (2.262)\frac{\sqrt{0.00011041}}{\sqrt{76{,}167.6363636}}$$
$$\approx -0.0013 \pm 0.0001 = [-0.0014, -0.0012]$$

is a 95% confidence interval for β_1. With 95% confidence, a one degree increase in temperature *decreases* the average peak height by at least 0.0012 mm and at most 0.0014 mm.

11.27 (a)

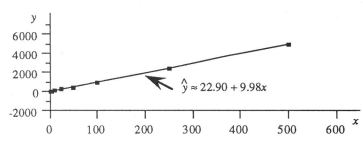

(b) $r^2 \approx 0.99995938 \approx 1.00$; so, almost all the variability in cost is associated with the linear model.

(c) $SSY = (n - 1) s_Y^2 \approx 7(3{,}057{,}369.6429) = 21{,}401{,}587.5003$
$r^2 \approx 0.99995938$ from (b)
$SSM = (SSY)r^2 \approx (21{,}401{,}587.5003)(0.99995938) \approx 21{,}400{,}718.1678$
$SSR = SSY - SSM \approx 21{,}401{,}587.5003 - 21{,}400{,}718.1678 = 869.3325$

(d) $MSR = (SSR)/(n - 2) \approx 869.3325/6 = 144.88875$

(e) $H_0: \beta_1 = 0$; $H_a: \beta_1 \neq 0$; Test statistic: $T = B_1\sqrt{SSX}/\sqrt{MSR}$; $\nu = 6$
Decision rule: Reject H_0 if the p-value is less than α.
(We will use $\alpha = 0.05$.)
Analysis: $SSX = 7s_X^2 \approx 7(30{,}723.6964285) = 215{,}065.8749995$. Since
$b_1 \approx 9.97535639$ and $MSR \approx 144.88875$, the observed value of T is
$t \approx (9.97535639)\sqrt{215{,}065.8749995}/\sqrt{144.88875} \approx 384.3$. The p-value
is $2P(T \geq 384.3) \approx 0.0000$. Reject H_0 and conclude that $\beta_1 \neq 0$.

(f) $b_1 \approx 9.97535639$, $SSX \approx 215{,}065.8749995$, and $MSR \approx 144.88875$.
The 97.5th quantile of a t distribution with 6 degrees of freedom is
2.447. So, $b_1 \pm (2.447)\frac{\sqrt{MSR}}{\sqrt{SSX}} \approx 9.97535639 \pm (2.447)\frac{\sqrt{144.88875}}{\sqrt{215{,}065.8749995}} \approx$
$9.97535639 \pm 0.06351315 \approx [9.91, 10.04]$ is a 95% confidence interval
for β_1. With 95% confidence, a 1 unit increase in lot size increases
the average cost by at least \$9.91 and at most \$10.04. Zero is not
included in this interval; so, as in (e), we conclude that $\beta_1 \neq 0$.

(g) Assumptions: The cost at each lot size is normally distributed.
The variability in cost is the same at each lot size. The costs at
different lot sizes are statistically independent.

11.29 $SSX = (n-1)s_X^2 \approx 12(15.16666667) \approx 182.000$, $b_1 \approx 7.478$ (Figure 11.4),
$SSR \approx 317.18$ (Example 11.8), $MSR = (SSR)/(n-2) \approx 317.18/11 \approx 28.835$,
and $T_{0.975} = 2.201$ for a t distribution with 11 degrees of freedom. Thus,
$b_1 \pm (2.201)\sqrt{MSR}/\sqrt{SSX} \approx 7.478 \pm (2.201)\sqrt{28.835}/\sqrt{182.000} \approx$
$7.478 \pm 0.876 = [6.602, 8.354] \approx [6.60, 8.35]$ is a 95% confidence interval for
β_1. With 95% confidence, an increase of 1 hour in the beating time tends
to increase the average Schopper-Riegler reading by at least 6.60
degrees and at most 8.35 degrees.

Section 11.4.1

11.31 (a)

Source	df	SS	MS	F-ratio	p-value
Linear model	1	40	40	10	0.0027
Residual	50	200	4		
Total	51	240			

(b) $H_0: \beta_1 = 0$; $H_a: \beta_1 \neq 0$
Test statistic: $F = (MSM)/(MSR)$ with $\nu_1 = 1$ and $\nu_2 = 50$
Decision rule: Reject H_0 if the p-value is less than $\alpha = 0.01$.
Analysis: The observed value of the test statistic is $f = 10$. The
p-value of the test is $P(F \geq 10) \approx 0.00266$. Since the p-value is less
than 0.01, reject H_0.

(c) Too little information is available to calculate SSX. Thus, a
confidence interval for β_1 cannot be calculated.

11.33 $H_0: \beta_1 = 0; H_a: \beta_1 \neq 0$

Test statistic: $F = (MSM)/(MSR)$ with $v_1 = 1$ and $v_2 = 6$

Decision rule: Reject H_0 if the p-value is less than α.

Analysis: From the solution of Problem 11.27, we find: $n = 8$, $SSM \approx 21,400,718.2$, $MSM = (SSM)/1 \approx 21,400,718.2$, $SSR \approx 869.3325$, $MSR = (SSR)/6 \approx 144.9$, and $SSY \approx 21,401,587.5$. Thus, $f \approx \frac{21,400,718.2}{144.9} \approx$ 147,693.0 and the p-value is $P(F > 147,693.0) \approx 0.0000$. H_0 can be rejected for any reasonable value of α. An *ANOVA* summary of this analysis is presented in the following table.

Source	df	SS	MS	F-ratio	p-value
Linear model	1	21,400,718.2	21,400,718.2	147,693.0	0.0000
Residual	6	869.3	144.9		
Total	7	21,401,587.5			

11.35 From the solution to Problem 11.30, $SSY = 50.5$ with 9 degrees of freedom, $SSM \approx 46.5993$ with 1 degree of freedom, and $SSR \approx 3.9007$ with 8 degrees of freedom. Using these results, the following *ANOVA* table is obtained. Since the p-value is so small, $H_0: \beta_1 = 0$ can be rejected for any reasonable value of α.

Source	df	SS	MS	F-ratio	p-value
Linear model	1	46.5993	46.5993	95.57	0.0000
Residual	8	3.9007	0.4876		
Total	9	50.5000			

11.37 **(a)** $\hat{y} \approx 11.9100 - 0.1668x$.

(b) $r^2 \approx 0.999971 \approx 1.00$. Almost all of the variability in the observed y values is associated with the least squares line.

(c)

Source	df	SS	MS	F-ratio	p-value
Linear model	1	289.4914	289.4914	> 1,000	0.0000
Residual	37	0.0085	0.0002		
Total	38	289.4999			

$H_0: \beta_1 = 0; H_a: \beta_1 \neq 0$
Test statistic: $F = (MSM)/(MSR)$ with $v_1 = 1$ and $v_2 = 37$
Decision rule: Reject H_0 if the p-value is less than α.
Analysis: From the preceding *ANOVA* summary, the observed value of the test statistic is $f \approx 289.4914/0.0002 = 1,447,457$. The p-value, $P(F > f) \approx 0.0000$, is so small that $H_0: \beta_1 = 0$ can be rejected for any reasonable value of α.

(d) $SSX = 13(20^2 + 40^2 + 60^2) - [13(20) + 13(40) + 13(60)]^2 / 39 = 10,400$
The 99.5th quantile of a t distribution with 37 degrees of freedom is 2.715, $MSR \approx 0.0002$, and $b_1 \approx -0.1668$. Thus,

$b_1 \pm (2.715)\sqrt{MSR}/\sqrt{SSX} \approx -0.1668 \pm 2.715\sqrt{0.0002}/\sqrt{10,400} \approx$
$-0.1668 \pm 0.0004 = [-0.1672, -0.1664]$ is a 99% confidence interval for
β_1. With 99% confidence, an increase of 1 degree in the spark
advance tends to decrease the average timing by at least 0.1664
msec and at most 0.1672 msec.

(e) Assumptions: The timing is normally distributed at each level of
the spark advance. The variability in timing is the same at each
level of the spark advance. The timings at different levels of the
spark advance are statistically independent.

Section 11.5

11.39 (a) Using $\hat{y} = 22.90 + 9.98x$, $22.90 + 9.98(200) = \$2,018.90$ is a point estimate of the
average cost for a lot of 200 units.

(b) Using $MSR = 144.888$; $x^* = 200$; $\hat{y}^* = 2,018.90$; $n = 8$, $\bar{x} = \Sigma x_i/8 =$
$941/8 = 117.625$; $SSX = 215,065.875$; $T_{0.975} = 2.447$, and the formula
given in Table 11.9,

$$\hat{y}^* \pm T_{0.975}\sqrt{MSR\left[\frac{1}{n} + \left(\left(x^* - \bar{x}\right)^2 \Big/ SSX\right)\right]} \approx 2018.90 \pm 2.447\sqrt{144.888\left(\frac{1}{8} + \frac{(200-117.625)^2}{215,065.875}\right)}$$

$$\approx 2,018.90 \pm 11.65 = [2007.25, 2030.55]$$

We are 95% confident that the true average cost of manufacturing
when the lot size is 200 units is between \$2007.25 and \$2030.55.

(c) Since $\hat{y} = 22.90 + 9.98x$, we predict that the cost of the next lot of
75 units will be $22.90 + 9.98(75) = \$771.40$. Using $x^* = 75$,
$\hat{y}^* = 771.40$, $T_{0.995} = 3.707$, $MSR = 144.888$, $\bar{x} = 117.625$,
$SSX = 215,065.875$, and the formula given in Table 11.12,

$$\hat{y}^* \pm T_{0.995}\sqrt{(MSR)\left(1 + \frac{1}{n} + \frac{(x^*-\bar{x})^2}{SSX}\right)} \approx 771.40 \pm 3.707\sqrt{144.888\left(1 + \frac{1}{8} + \frac{(75-117.625)^2}{215,065.875}\right)}$$

$$\approx 771.40 \pm 47.50 \approx [723.90, 818.90]$$

is a 99% prediction interval for the cost of producing the next lot of
75 items.

11.41 From Problem 11.17(a), $\bar{x} = \Sigma x_i/6 = 21/6 = 3.5$, $\hat{y} = \frac{17}{15} + \frac{507}{35}x$, $MSR =$
$\frac{1646}{105}$, and $SSX = 17.5$. For a t distribution with 4 degrees of freedom, the
97.5th quantile is 2.776.

(a) Using the formula in Table 11.9,

$$\hat{y}^* \pm T_{0.975}\sqrt{(MSR)\left(\frac{1}{n} + \frac{(x^*-\bar{x})^2}{SSX}\right)} \approx \left(\frac{17}{15} + \frac{507}{35}x^*\right) \pm 2.776\sqrt{\frac{1646}{105}\left(\frac{1}{6} + \frac{(x^*-3.5)^2}{17.5}\right)}$$

is a 95% confidence interval for $E[Y \mid x = x^*]$. Rounding to 4
decimal places, we obtain $51.8333 \pm 4.4871 = [47.3462, 56.3204]$

when $x = 3.5$. When $x = 4.5$, $66.3190 \pm 5.1997 = [61.1193, 71.5187]$ is obtained.

The other 6 confidence intervals, obtained using JMP, are summarized in the following table. If the preceding formula is used, slight differences in answers will be noted. Those are due to rounding.

x	y	Lower Conf	Upper Conf	Lower Pred	Upper Pred
1	14	7.6631	23.5750	2.0494	29.1887
2	33	24.1318	36.0777	17.5942	42.6154
3	40	39.9144	49.2666	32.6445	56.5364
4	63	54.4001	63.7523	47.1303	71.0221
5	76	67.5889	79.5349	61.0513	86.0725
6	85	80.0917	96.0035	74.4780	101.6173

(b)

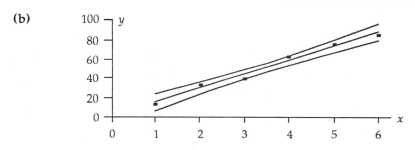

(c) Using the formula in Table 11.12, a 95% prediction interval for the next Y at $x = x^*$ is

$$\hat{y}^* \pm T_{0.975}\sqrt{(MSR)\left(1 + \tfrac{1}{n} + \frac{\left(x^*-\bar{x}\right)^2}{SSX}\right)} \approx \left(\tfrac{17}{15} + \tfrac{507}{35}x^*\right) \pm 2.776\sqrt{\tfrac{1646}{105}\left(1 + \tfrac{1}{6} + \frac{(x^*-3.5)^2}{17.5}\right)}.$$

Rounding to 4 decimal places, we obtain $51.8333 \pm 11.8717 = [39.9616, 63.7050]$ when $x = 3.5$. When $x = 4.5$, $66.3190 \pm 12.1590 = [54.1600, 78.4780]$ is obtained. The other 6 prediction intervals, obtained using JMP, are summarized in the table in (a).

(d)

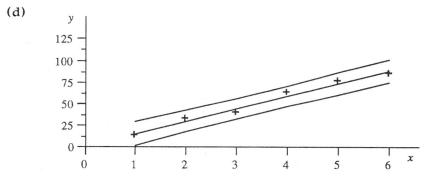

135

11.43 From discussions and examples in Sections 11.2 and 11.3, $n = 15$, $\hat{y} = 15.5 - 1.1x$, $\bar{x} = 5$, $SSX = 30$, and $MSR \approx 1.21$. The 95th percentile of a t distribution with 13 degrees of freedom is 1.771.

(a) A 90% confidence interval for the average coating thickness at a pressure of x pounds is given by

$$(15.5 - 1.1x) \pm 1.771\sqrt{1.21\left(\tfrac{1}{15} + \tfrac{(x-5)^2}{30}\right)}.$$

Interval endpoints at selected values of x are summarized in the following table.

x	$\hat{y}$	Lower conf	Upper conf	Lower pred	Upper pred
3.0	12.20	11.329	13.071	10.066	14.334
4.0	11.10	10.484	11.716	9.057	13.143
4.5	10.55	10.016	11.084	8.530	12.570
5.0	10.00	9.497	10.503	7.988	12.012
5.5	9.45	8.916	9.984	7.430	11.470
6.0	8.90	8.284	9.516	6.857	10.943
7.0	7.80	6.929	8.671	5.666	9.934

(b) A 90% confidence band is included in the following JMP graphic. The band widens as the absolute deviation of x from $\bar{x}$ is increased. That is, the minimum width of a 90% confidence interval for the average coating thickness occurs at $x = \bar{x} = 5$.

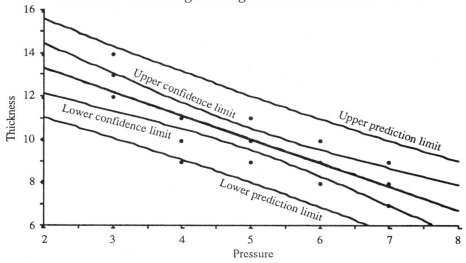

(c) A 90% prediction interval for the average coating thickness of the next substrate produced at a pressure of x pounds is given by

$$(15.5 - 1.1x) \pm 1.771\sqrt{1.21\left(1 + \tfrac{1}{15} + \tfrac{(x-5)^2}{30}\right)}.$$

See the table in (a) for interval endpoints at selected values of x.

(d) A 90% prediction band is included in the graphic in (b).

136

11.45 **(a)** $\hat{y} = -9.4020 + 2.8445x$

(b) A scatter plot with the least squares line included follows. It appears that use of a simple linear model is reasonable.

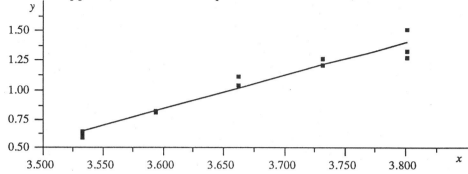

(c) $r^2 \approx 0.9515$. Approximately 95.15% of the variability in solubility is associated with the least squares line.

(d)

Source	df	SS	MS	F-ratio	p-value
Linear model	1	1.0264	1.0264	197.4	0.0000
Residual	10	0.0524	0.0052		
Total	11	1.0788			

Since the p-value (0.0000) of the tabled F test is so small, $H_0 : \beta_1 = 0$ can be rejected for any reasonable value of α.

[*Note:* $\hat{y} = -9.4020 + 2.8445x$, $b_1 = 2.8445$, $n = 12$, $\bar{x} = 3.6645$, $SSX \approx 0.1269$, and $MSR \approx 0.0052$ are used to answer (e), (f), and (g). Also, for a t distribution with 10 degrees of freedom, $T_{0.95} = 1.812$, $T_{0.975} = 2.228$ and $T_{0.995} = 3.169$.]

(e)
$$b_1 \pm T_{0.995} \sqrt{MSR/SSX} \approx 2.8445 \pm 3.169\sqrt{0.0052/0.1269}$$
$$\approx 2.8445 \pm 0.6415 = [2.2030, 3.4860]$$

We are 99% confident that a 1 unit increase in reciprocal temperature will increase the average solubility at least 2.2030 units and at most 3.4860 units. Zero is not included in this interval; so, as in (d), we conclude that there is a linear relationship between reciprocal temperature and solubility.

(f) Since $\hat{y} = -9.4020 + (2.8445)(3.6) = 0.8382$ at $x = 3.6$,

$$0.8382 \pm T_{0.95}\sqrt{(MSR)\left(\frac{1}{n} + \frac{(3.6-\bar{x})^2}{SSX}\right)}$$
$$\approx 0.8382 \pm 1.812\sqrt{0.0052\left(\frac{1}{12} + \frac{(3.6-3.6645)^2}{0.1269}\right)}$$
$$\approx 0.8382 \pm 0.0445 = [0.7937, 0.8827]$$

137

is a 90% confidence interval for the average solubility at a reciprocal temperature of 3.6. That is, the average solubility at x = 3.6 is at least 0.7937 and at most 0.8827 (with 90% confidence).

(g) Since $\hat{y}$ = −9.4020 + (2.8445)(3.7) = 1.12265 ≈ 1.1226 at x = 3.7,

$$1.1226 \pm T_{0.975}\sqrt{(MSR)\left(1 + \frac{1}{n} + \frac{(3.7-\bar{x})^2}{SSX}\right)}$$

$$\approx 1.1226 \pm 2.228\sqrt{0.0052\left(1 + \frac{1}{12} + \frac{(3.7-3.6645)^2}{0.1269}\right)}$$

$$\approx 1.1226 \pm 0.1680 = [0.9546, 1.2906]$$

is a 95% prediction interval for the solubility when the reciprocal temperature is 3.7. That is, the solubility of nitrous oxide in nitrogen dioxide will be at least 0.9546 percent by weight and at most 1.2906 percent by weight, the next time a reciprocal temperature of 3.7 is used.

(h) Assumptions: The solubility at each reciprocal temperature level is normally distributed. The variability in the solubility is the same at each reciprocal temperature level. The solubilities at different reciprocal temperatures are statistically independent.

Section 11.6

11.47 (a)

Degrees	j	s_j^2	n_j	$n_j - 1$	$(n_j - 1)s_j^2$
20	1	0.000351808	13	12	0.004221696
40	2	0.000243744	13	12	0.002924928
60	3	0.000110526	13	12	0.001326312
Total		Error degrees of freedom = 36			0.008472936 (SSE)

$H_0: E[Y \mid x] = \beta_0 + \beta_1 x$
H_a: "A model other than the simple linear model should be used."
Test statistic: $F = (MSLOF)/(MSE)$; $v_1 = 1$ and $v_2 = 36$
Decision rule: Reject H_0 if the p-value is less than or equal to α. The p-value is $P(F \ge f)$, with f the observed value of the test statistic.
Analysis: $SSY \approx 289.49993667 \approx 289.499937$; $r^2 \approx 0.999970599$
SSM = $(SSY)r^2$
$\approx (289.49993667)(0.999970599) \approx 289.491425$
SSR = $(SSY)(1 - r^2)$
$\approx (289.49993667)(1 - 0.999970599) \approx 0.008512$
$SSE \approx 0.008473$ from the preceding table.
$MSE = SSE/36 \approx 0.008473/36 \approx 0.000235$
$SSLOF = SSR - SSE \approx 0.008512 - 0.008473 = 0.000039$

$MSLOF = SSLOF/1 = SSLOF \approx 0.000039$
Thus, $f = MSLOF/MSE \approx 0.000039/0.000235 \approx 0.1660$
and the p-value is $P(F \geq 0.1660) \approx 0.6861$. There is
insufficient evidence to reject the simple linear model
in favor of some other model.

(b)

Source	df	SS	MS	F-ratio	p-value
Linear model	1	289.491425	289.491425	> 1000	0.0000
Residual	37	0.008512	0.000230		
Lack of fit	1	0.000039	0.000039	0.1660	0.6861
Pure error	36	0.008473	0.000235		
Total	38	289.499937			

11.49 (a)

Casting weight	j	s_j^2	n_j	$n_j - 1$	$(n_j - 1)s_j^2$
2.590	1	0.00020000	2	1	0.00020000
2.615	2	0.00002500	3	2	0.00005000
2.625	3	0.00005000	2	1	0.00005000
2.650	4	0.00001250	2	1	0.00001250
2.655	5	0.00000000	2	1	0.00000000
Total		Error degrees of freedom = 6			0.00031250 (SSE)

Rough casting weight (x) values at which at least 2 finished rod
weights (Y) were observed are given in the preceding table. Since
$SSE = 0.00031250$ from that table and $SSR \approx 0.00297763$ from
Problem 11.36, $SSLOF = SSR - SSE \approx 0.00266513$. Adding this
information to the ANOVA summary in Problem 11.36 produces
the following table.

Source	df	SS	MS	F-ratio	p-value
Linear model	1	0.01415837	0.014158	109.75	0.0000
Residual	23	0.00297763	0.000129		
Lack of fit	17	0.00266513	0.000157	3.02	0.0888
Pure error	6	0.00031250	0.000052		
Total	24	0.01713600			

(b) $H_0: E[Y \mid x] = \beta_0 + \beta_1 x$
H_a: "A model other than the simple linear model should be used."
Test statistic: $F = (MSLOF)/(MSE)$; $v_1 = 17$ and $v_2 = 6$
Decision rule: Reject H_0 if the p-value is less than or equal to α.
        The p-value is $P(F \geq f)$, with f the observed value
        of the test statistic.

Analysis: From the table in (a), $f = MSLOF/MSE \approx$
    $0.000157/0.000052 \approx 3.02$ and the p-value is
    approximately $P(F \geq 3.02) \approx 0.0888$. The simple linear
    model can be rejected as adequate for any $\alpha \geq 0.0888$.

11.51 (a) $\hat{y} \approx 125.579 + 11.188x$

(b) Based on the scatter plot that follows, use of a simple linear model is reasonable.

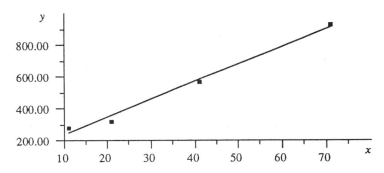

(c) $r^2 \approx 0.9894$. The least squares line accounts for approximately 98.94% of the variability in the sample of y values.

(d)

Source	df	SS	MS	F-ratio	p-value
Linear model	1	1,314,249.8853	1,314,249.8853	1,683.2	0.0000
Residual	18	14,054.8263	780.8236		
Total	19	1,328,304.7116			

$H_0: \beta_1 = 0; H_a: \beta_1 \neq 0$
Test statistic: $F = (MSM)/(MSR)$ with $\nu_1 = 1$ and $\nu_2 = 18$
Decision rule: Reject H_0 if the p-value is less than α.
Analysis: From the preceding *ANOVA* summary, the observed
value of the test statistic is $f \approx 1,314,249.8853/780.8236$
$\approx 1,683.2$. The p-value, $P(F > f) \approx 0.0000$, is so small
that $H_0: \beta_1 = 0$ can be rejected for any reasonable value
of α.

[Note: The sums of squares in the preceding table were obtained
using JMP. The sums of squares in the text answer were obtained
using StataQuest. The differences are due to round-off error.]

(e)

Source	df	SS	MS	F-ratio	p-value
Model	1	1,314,249.8853	1,314,249.8853	1,683.2	0.0000
Residual	18	14,054.8263	780.8236		
Lack of fit	2	14,014.9550	7,007.4775	2,812.0	0.0000
Pure error	16	39.8713	2.4920		
Total	19	1,328,304.7116			

$H_0: E[Y \mid x] = \beta_0 + \beta_1 x$
$H_a:$ "A model other than the simple linear model should be used."
Test statistic: $F = (MSLOF)/(MSE); \nu_1 = 2$ and $\nu_2 = 16$
Decision rule: Reject H_0 if the p-value is less than or equal to α.

140

Analysis: From the table in (a), $f = MSLOF/MSE \approx 2812.0$ is observed and the p-value is approximately $P(F \geq 2812.0) \approx 0.0000$. We conclude that a model other than the simple linear model may be more appropriate.

[Note: A significant lack of fit does not mean we should go out of our way to find another model. The coefficient of determination (0.9894) is so large that the simple linear model may provide the investigators the needed information.]

Section 11.7.2

11.53 (a) Using MYSTAT, the following tabled values were calculated to the nearest thousandth.

x	y	y-hat	residual	studentized residual
1.0	13	13.186	−0.186	−0.179
1.5	14	13.868	0.132	0.119
2.0	14	14.550	−0.550	−0.484
2.5	16	15.232	0.768	0.668
3.5	15	16.596	−1.596	−1.561
4.0	18	17.278	0.722	0.619
4.5	20	17.960	2.040	2.447
5.5	18	19.324	−1.324	−1.409
6.0	20	20.006	−0.006	−0.006

(b) A plot of the studentized residuals versus the x values tends to "fan out" as the values of x are increased. Since the plotted points do not form a random pattern, at least one of the model assumptions is likely to be false.

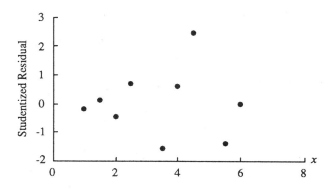

11.55 (a) Using MYSTAT, the following tabled values were calculated to the nearest thousandth.

x	y	y-hat	residual	studentized residual
-92	0.375	0.375	0.000	0.043
-50	0.331	0.321	0.010	1.107
-40	0.300	0.308	-0.008	-0.818
-3	0.251	0.260	-0.009	-0.956
23	0.224	0.227	-0.003	-0.298
40	0.216	0.205	0.011	1.073
52	0.204	0.190	0.014	1.496
78	0.146	0.157	-0.011	-1.079
125	0.081	0.096	-0.015	-1.802
162	0.056	0.049	0.007	0.777
176	0.035	0.031	0.004	0.448

(b) The points in the following plot of the studentized residuals versus x form a random pattern. (The same is true of a plot of the studentized residuals versus the estimated values.)

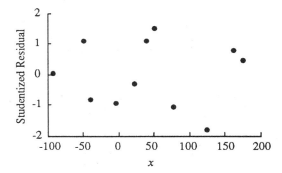

Since the normal quantile plot of the residuals is reasonably linear and the p-value of the Shapiro-Wilk W test is moderately large, an assumption of normality seems reasonable.

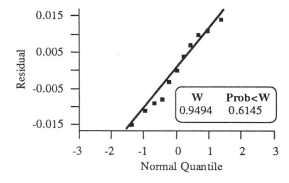

142

11.59 (a)

x	y	y-hat	residual	standardized residual
20	8.533	8.573141	−0.040140	−2.735690
20	8.591	8.573141	0.017859	1.217124
20	8.560	8.573141	−0.013140	−0.895590
20	8.590	8.573141	0.016859	1.148972
20	8.575	8.573141	0.001859	0.126693
20	8.560	8.573141	−0.013140	−0.895590
20	8.582	8.573141	0.008859	0.603757
20	8.571	8.573141	−0.002140	−0.145920
20	8.568	8.573141	−0.005140	−0.350370
20	8.557	8.573141	−0.016140	-1.100040
20	8.576	8.573141	0.002859	0.194845
20	8.599	8.573141	0.025859	1.762340
20	8.598	8.573141	0.024859	1.694188
40	5.199	5.236333	−0.037330	−2.493620
40	5.248	5.236333	0.011667	0.779255
40	5.225	5.236333	-0.011330	−0.756990
40	5.238	5.236333	0.001667	0.111322
40	5.224	5.236333	−0.012330	−0.823780
40	5.241	5.236333	0.004667	0.311702
40	5.243	5.236333	0.006667	0.445289
40	5.261	5.236333	0.024667	1.647569
40	5.225	5.236333	−0.011330	−0.756990
40	5.228	5.236333	−0.008330	−0.556610
40	5.248	5.236333	0.011667	0.779255
40	5.245	5.236333	0.008667	0.578876
40	5.229	5.236333	−0.007330	−0.489820
60	1.879	1.899526	−0.020530	−1.398860
60	1.897	1.899526	−0.002530	−0.172130
60	1.887	1.899526	−0.012530	−0.853650
60	1.892	1.899526	−0.007530	−0.512890
60	1.911	1.899526	0.011474	0.782000
60	1.896	1.899526	−0.003530	−0.240280
60	1.900	1.899526	0.000474	0.032329
60	1.911	1.899526	0.011474	0.782000
60	1.912	1.899526	0.012474	0.850152
60	1.909	1.899526	0.009474	0.645696
60	1.910	1.899526	0.010474	0.713848
60	1.905	1.899526	0.005474	0.373088
60	1.894	1.899526	−0.005530	−0.376580

(b) Standardized residuals for the 1st and 14th observations have
absolute values of approximately 2.7 and 2.5, respectively. A
study of the circumstances at the times the observations were
obtained should be made to determine whether they are
representative of the true situation or are in error.

The following plot of the standardized residual versus spark
advance seems to indicate that the variability in timing
decreases as spark advance increases. This impression is largely
due to the size of the residuals for the 1st and 14th observations.

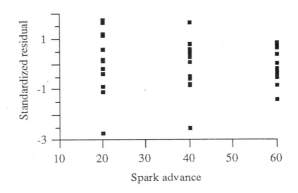

The standardized residuals for the 1st and 14th observations are well off a line through the main body of the following normal quantile plot of the standardized residuals. This reinforces the impression that those observations may be due to some special cause that was present at the time the data were obtained.

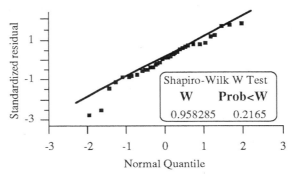

11.61 (a)

$$\hat{y} = 0.266 + 0.002x$$

(b) The following normal quantile plot of the standardized residuals and the Shapiro-Wilk W test support an assumption of normality.

The wedge-shaped pattern in the following plot of the standardized residuals versus mileage probably indicates that the variance of emissions increases as the mileage increases.

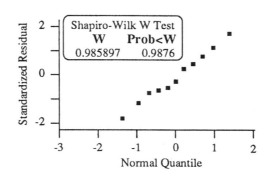

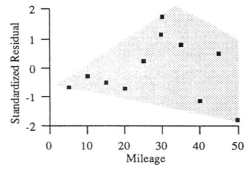

Section 11.7.3

11.63 **(a)** For $y' = \ln(y)$ and $x' = \ln(x)$, $\hat{y}' \approx 1.9558x' - 15.6261$.

(b) The scatter plot of the transformed data has a linear pattern. Most of the points tend to follow the least squares line. However, if the two deviant points are deleted, fitting a third degree polynomial to the remaining points seems more appropriate.

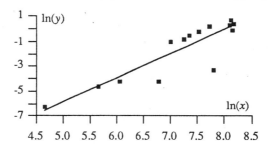

(c) The following plot of the standardized residual versus $\ln(x)$ indicates an increase in variability as x increases or a lack of fit.

The following normal quantile plot and Shapiro-Wilk W test indicate that the residuals have a nonnormal distribution.

145

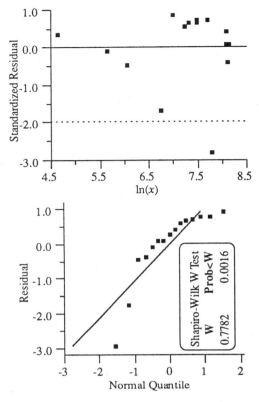

(d) The fit, though far from perfect, may prove useful for point estimation and prediction. The least squares line for this model fits the data better than that of Problem 11.9. In this case, $r^2 = 0.78223$. For the simple linear model, $r^2 = 0.695837$.

Section 11.8.1

11.65 (a) Consider the following JMP output. The constant, coefficient of x, and coefficient of x^2 for a least squares polynomial of degree 2 are given in the Intercept, x, and $x*x$ rows of the estimate column. The least squares polynomial of degree 2 is

$$\hat{y} \approx 0.0014x^2 - 0.0879x + 11.5595.$$

Parameter Estimates				
Term	**Estimate**	**Std Error**	**t Ratio**	**Prob>|t|**
Intercept	11.559506	0.86994	13.29	0.0000
x	-0.087941	0.03555	-2.47	0.0242
x*x	0.0014424	0.0003	4.77	0.0002

146

(b)   From the following JMP summary, $R^2 \approx 0.90$. Approximately 90% of the variability in the y values is associated with the quadratic model.

Summary of Fit	
RSquare	0.9011
RSquare Adj	0.8895
Root Mean Square Error	0.9104
Mean of Response	13.4250
Observations	20.0000

The points in the following scatter plot tend to fall along the fitted polynomial. The fit is a good one.

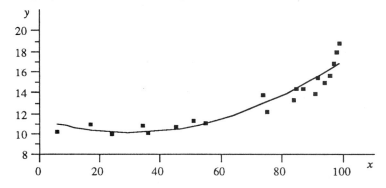

(c)   Consider the following plots of the residuals. The Shapiro-Wilk W test (p-value ≈ 0.7970) and the normal quantile plot support an assumption of normality. The plot of the residuals versus the predicted values indicates a possible lack of fit condition at the higher values. It appears that statistical inferences will be valid except, possibly, at large values of x.

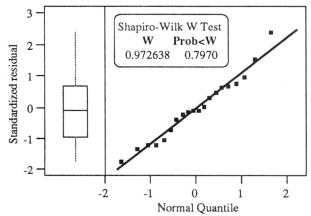

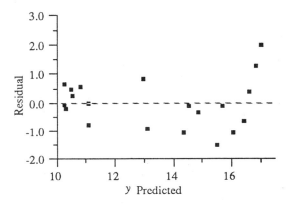

11.67 **(a)** A scatter plot is presented in (c). The plotted points take on a parabolic shape. A quadratic model is appropriate.

(b) Consider the following JMP output. The constant, coefficient of x, and coefficient of x^2 for a least squares polynomial of degree 2 are given in the Intercept, x, and x^2 rows of the estimate column. The least squares polynomial of degree 2 is $\hat{y} = 11 - 2x + 0.1x^2$.

Parameter Estimates				
Term	Estimate	Std Error	t Ratio	Prob>|t|
Intercept	11	4.459e-7	2.47e7	0.0000
x	-2	7.379e-8	-2.7e7	0.0000
x^2	0.1	2.777e-9	3.6e+7	0.0000

(c)

Analysis of Variance				
Source	DF	Sum of Squares	Mean Square	F Ratio
Model	2	259.31429	129.657	1.255e15
Error	11	0.00000	0.000	Prob>F
C Total	13	259.31429		0.0000

The fit is a perfect one. SSM = SSY, so there is no residual error. The model should provide excellent point estimates.

148

11.71 **(a)** Consider the following Minitab output. Using the x3 row, the p-value of a test of $H_0: \beta_3 = 0$ versus $H_a: \beta_3 \neq 0$ is 0.344. Thus, there is insufficient evidence to reject H_0. This does not mean that β_3 should be deleted from the model. However, further considerations may lead to the exclusion of β_3.

Regression Analysis

The regression equation is
y = - 39.9 + 0.716 x1 + 1.30 x2 - 0.152 x3

Predictor	Coef	Stdev	t-ratio	p
Constant	-39.92	11.90	-3.36	0.004
x1	0.7156	0.1349	5.31	0.000
x2	1.2953	0.3680	3.52	0.003
x3	-0.1521	0.1563	-0.97	0.344

s = 3.243   R-sq = 91.4%   R-sq(adj) = 89.8%

(b) Values in the Coef column of the following Minitab summary are $b_0 \approx -50.359$, $b_1 \approx 0.671$, and $b_2 \approx 1.295$. Thus,

$$\hat{y} \approx - 50.359 + 0.671x_1 + 1.295x_2 .$$

Regression Analysis

The regression equation is
y = - 50.4 + 0.671 x1 + 1.30 x2

Predictor	Coef	Stdev	t-ratio	p
Constant	-50.359	5.138	-9.80	0.000
x1	0.6712	0.1267	5.30	0.000
x2	1.2954	0.3675	3.52	0.002

s = 3.239   R-sq = 90.9%   R-sq(adj) = 89.9%

(c) Using the Minitab summaries in (a) and (b), the adjusted coefficients of determination for the full and reduced models are 89.8% and 89.9%, respectively. The similarity in the two results supports use of the reduced model.

(d)

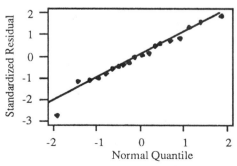

The preceding Minitab normal quantile plot is reasonably linear. The standardized residual (-2.73221) associated with the 21st observation lies off a line through the main body of the plot and should be investigated as a possible outlier. Using JMP, the Shapiro-Wilk W test (p-value = 0.7113) also supports a normality assumption.

The following Minitab plot of the standardized residuals versus the fitted values gives a slight indication of unequal variances. However, much of that impression is due to the 21st observation, which has a standardized residual of -2.73221.

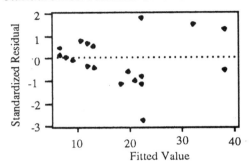

The ($\hat{y}, y$) pairs in the following plot fall along the line $y = \hat{y}$, indicating that the assumed model equation is a reasonable one.

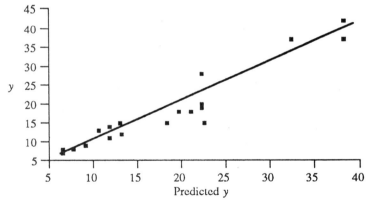

(e) Based on the evidence presented in (c) and (d), the reduced model seems to describe the data as well as the full model.

11.73 Using $x_1 = 4.0$, $x_2 = 1.5$, and $1 - \alpha = 0.95$, the following Minitab output results. With 95% confidence, the damage incurred when a 4000 pound shipment is moved 1500 miles is at least $126.87 and at most $179.61.

Fit	Stdev.Fit	95.0% C.I.	95.0% P.I.
153.24	8.28	(126.87, 179.61)	(96.14, 210.34)

Using Minitab with $x_1 = 3.0$, $x_2 = 2.2$, and $1 - \alpha = 0.90$, the following output results. Thus, [$86.04, $117.25] is a 90% prediction interval for the amount of damage incurred the next time a 3000 pound shipment is moved 2200 miles.

Fit	Stdev.Fit	90.0% C.I.	90.0% P.I.
1131.64	11.05	(105.63, 157.66)	(86.04, 177.25)

Section 11.9

11.75 The following normal quantile plot of the standardized residuals, which is reasonably linear, and the Shapiro-Wilk W test (p-value = 0.3130) support an assumption of normality.

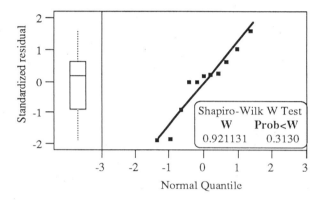

The following plot of the standardized residuals versus the predicted value provides some evidence of curvature. The points are not randomly scattered about the horizontal dashed line. However, the coefficient of determination ($r^2 \approx 0.967612$) indicates that the model accounts for approximately 96.8% of the variability in the $\ln(y)$ values. An assumption that the least squares line adequately describes the relationship between $\ln(y)$ and $\ln(x)$ seems reasonable.

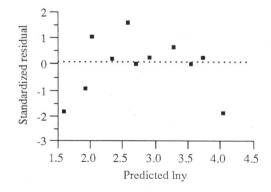

151

The following plot of the standardized residuals versus ln(x) does not provide strong evidence against an assumption of equal variances. It does provide some evidence of curvature, but we have decided that the curvature is not severe enough to cause serious deficiencies when the linear model is used.

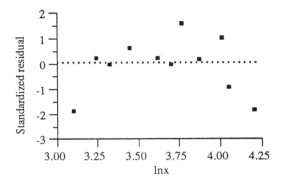

11.76 The following normal quantile plot of the standardized residuals has distinct curvature. The *p*-value of the Shapiro-Wilk *W* test is quite small. These two observations seem to indicate that normality should not be assumed.

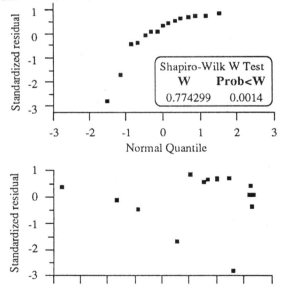

The preceding plot of the standardized residuals versus the natural logarithm of the runoff volume indicates that the peak discharge varies more at the higher values of the runoff volume.

The most extreme point on that plot is associated with observation 7. The point associated with that observation is the rightmost point that is well off the least squares line in the solution of Problem 11.63(b). This provides some indication of a nonlinear relationship between $\ln(x)$ and $\ln(y)$.

11.77 **(a)** Let $y' = \ln(y)$, with y the ordinate tabled in Problem 11.10. The least squares line for the transformed data is $\hat{y}' \approx 0.0059x + 2.1806$.

Parameter Estimates				
Term	Estimate	Std Error	t Ratio	Prob>|t|
Intercept	2.1806249	0.04798	45.44	0.0000
x	0.0059288	0.00065	9.06	0.0000

(b) Even though the adjusted $r^2 \approx 0.81$, the least squares line found in (a) is not a good fit to the data.

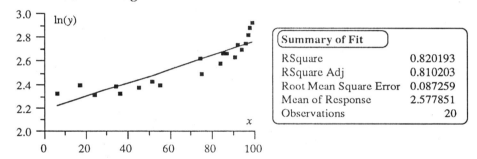

Summary of Fit	
RSquare	0.820193
RSquare Adj	0.810203
Root Mean Square Error	0.087259
Mean of Response	2.577851
Observations	20

(c) The following plot of the standardized residuals versus x has a U-shape. This may indicate that at least one of the model assumptions is not satisfied. It may also be due to the curvature (lack of fit) noted in (b).

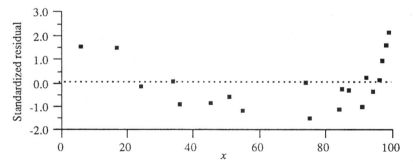

The following normal quantile plot of the standardized residuals exhibits some upward curvature. The p-value of the Shapiro-Wilk W test is 0.1484, so a normality assumption may be adequately satisfied. Combining the fact that the p-value of the

153

formal test is neither small nor moderately large and several points fall well off a line through the main body of points in a normal quantile point, an assumption of normality is questionable.

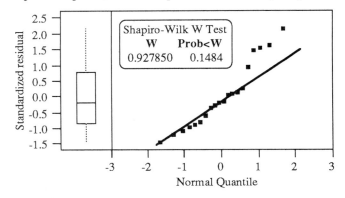

Supplementary Problems

11.87 From Problems 11.4 and 11.16, $n = 9$, $b_0 \approx 11.823$, $\sum_{i=1}^{9} x_i = 30.5$, and $MSR \approx 1.418$. Since $\bar{x} = 30.5/9 \approx 3.389$, $SSX = 8s_x^2 \approx 24.889$, and the 97.5th quantile of a t distribution with 7 degrees of freedom is 2.365,

$$b_0 \pm T_{0.975}\sqrt{(MSR)\left(\frac{1}{n} + \frac{\bar{x}^2}{SSX}\right)} \approx 11.823 \pm 2.365\sqrt{1.418\left(\frac{1}{9} + \frac{3.389^2}{24.889}\right)}$$

$$\approx 11.823 \pm 2.131 = [9.692, 13.954]$$

determines a 95% confidence interval for β_0. We are 95% confident that β_0 is at least 9.692 and at most 13.954. With 95% confidence, the average number of defective items produced during the time interval $(0, t)$ is at least 9.692 and at most 13.954.

11.89 Since $b_0 = \bar{y} - b_1\bar{x}$, $\hat{y} = b_0 + b_1 x = \left(\bar{y} - b_1\bar{x}\right) + b_1 x = \bar{y} + b_1(x - \bar{x})$. When $x = \bar{x}$, $\hat{y} = \bar{y} + b_1(\bar{x} - \bar{x}) = \bar{y}$. Therefore, $(\bar{x}, \bar{y})$ is on the least squares line.

154

CHAPTER 12
DESIGN AND ANALYSIS OF SINGLE-FACTOR EXPERIMENTS

Section 12.1.1

12.1

Electrolyte (i)	1	2	3	4	5
$\bar{y}_i$	45	40	40	41	34
s_i^2	14	8	$18\frac{2}{3}$	2	$\frac{2}{3}$

(a) $k = 5$; $n = 4$; $\bar{y} = (45 + 40 + 40 + 41 + 34)/5 = 40$

$s_{\text{pooled}}^2 = \left(14 + 8 + 18\frac{2}{3} + 2 + \frac{2}{3}\right)/5 = 26/3$; $s_{\text{pooled}} = \sqrt{26/3} \approx 2.94$

$\sqrt{(5-1)/(5 \times 4)} = \sqrt{0.2} \approx 0.45$

For $k = 5$, $v = 5(4 - 1) = 15$, and $\alpha = 0.05$, $h = 2.88$ from Appendix R.

Thus, $[\text{LDL, UDL}] = \bar{y} \pm h s_{\text{pooled}} \sqrt{(k-1)/(kn)}$

$\approx 40 \pm (2.88)(2.94)(0.45)$

$\approx 40 \pm 3.8 = [36.2, 43.8]$.

The associated graphical representation follows. Since $\bar{y}_1$ is greater than UDL (or $\bar{y}_5$ is less than LDL), we can reject equality of the population means.

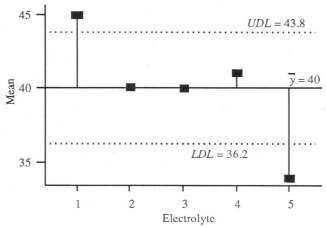

(**Note:** Minitab contains an ANOM option that will calculate the decision limits and plot the associated graphic. Decision limits are calculated using an estimate of σ based on the average range rather than the pooled variance. In this situation, those limits are LDL $\approx$ 35.77 and UDL $\approx$ 44.23.)

(b) The following scatter plot seems to indicate that (1) the population variances may not be equal, (2) electrolyte 1 has the largest population mean, (3) electrolyte 5 has the smallest population mean, and (4) the other population means may be equal. The range chart does not support unequal variances.

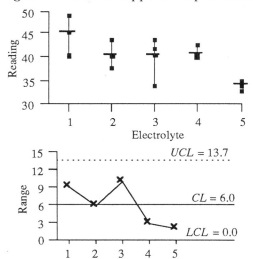

12.3

Supplier (i)	A = 1	B = 2	C = 3	D = 4	E = 5	Total
$\bar{y}_i$	99	100	103	98	100	500
s_i^2	1	1	1	1	1	5

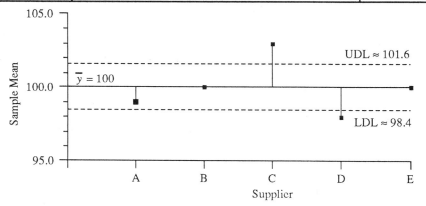

$k = 5; n = 3; \bar{\bar{y}} = 500/5 = 100; s_{pooled}^2 = 5/5 = 1; s_{pooled} = \sqrt{1} = 1$

$\sqrt{(k-1)/(kn)} = \sqrt{(5-1)/(5 \times 3)} = \sqrt{4/15} \approx 0.52$

For $k = 5$, $\nu = 5(3-1) = 10$, and $\alpha = 0.05$, $h = 3.07$ from Appendix R.

156

$$[\text{LDL, UDL}] = \bar{y} \pm hs_{\text{pooled}} \sqrt{(k-1)/(kn)} \approx 100 \pm (3.07)(1)(0.52) \approx 100 \pm 1.6 =$$
[98.4, 101.6]. Since at least one sample mean falls outside the decision interval, reject equality of the true average resistances.

Section 12.1.2

12.5 **(a)** Consider the following scatter plot. It appears that the average voltage for the Supplier D diodes is shifted the greatest amount. It appears that the shifts are more variable for the Supplier F diodes.

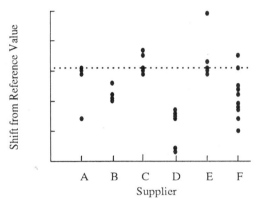

Several outliers are indicated on the following grouped box plot. Since their existence can adversely affect any formal analyses, they should be investigated thoroughly. The extreme deviations may be a result of measurement error, the order in which the measurements were obtained, or nonnormality.

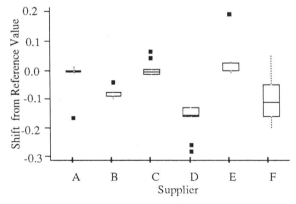

(b) The *p*-value (0.0000) in the following ANOVA summary is extremely small. Thus, we reject equality of means and conclude that at least 2 means differ.

Source	df	SS	MS	F-ratio	p-value
Supplier	5	0.3108	0.0622	25.9167	0.0000
Error	66	0.1556	0.0024		
Total	71	0.4664			

(**Note:** MYSTAT warns that -0.16 for Supplier A, 0.19 for Supplier E, and 0.05 for Supplier F are outliers. Their studentized residuals (Section 12.1.6) are -3.375, 4.039, and 3.420, respectively. They should be investigated further. Their existence could adversely affect any decisions made.)

(c) The standard deviation for the Supplier F data is outside the control limits on the following s chart, indicating unequal variances. Combining this with the observations in (a) and (b), assumptions of normality and equal variances are not justified.

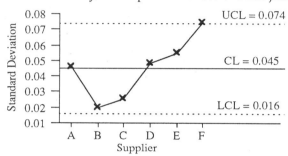

12.7

Source	df	SS	MS	F-ratio	p-value
Electrolyte	4	248.00	62.00	7.15	0.0020
Error	15	130.00	8.67		
Total	19	378.00			

The p-value of the F test for equal population means, summarized in the preceding ANOVA table, is 0.0020. Thus, the hypothesis of equal means can be rejected for any $\alpha \geq 0.0020$.

12.9 (a)

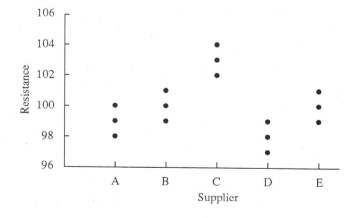

The preceding scatter plot supports an assumption of equal variances. Also, the plot suggests that the true average resistance for Supplier C is greater than that for Supplier D.

(b)

Source	df	SS	MS	F-ratio	p-value
Supplier	4	42.00	10.50	10.50	0.001
Error	10	10.00	1.00		
Total	14	52.00			

The p-value of the F test is 0.001. Since $\alpha = 0.05$ exceeds that value, there is sufficient evidence to reject equality of means. We conclude that the average resistances of at least two suppliers differ.

Section 12.1.3

12.11 (a)

Source	df	SS	MS	F-ratio	p-value
Coating	3	1135.0	378.3	29.8	0.0000
Error	16	203.2	12.7		
Total	19	1338.2			

(b) For $k = 4$, $v = 16$, and $p = 1 - \alpha = 0.95$, $Q_{0.95} = 4.05$ from Appendix S. Since $MSE = 12.7$ and $n = 5$, $w = 4.05\sqrt{12.7/5} \approx 6.45$. In the summary below, sample means that differ by at most 6.45 are indicated with the same grouping symbol. We conclude that $\mu_1 > \mu_3$, $\mu_2 > \mu_3$, $\mu_1 > \mu_4$, and $\mu_2 > \mu_4$. There is insufficient evidence for further claims.

Total	n	Mean	Coating	Grouping	
210	5	42.0	4	A	
218	5	43.6	3	A	
286	5	57.2	2		B
292	5	58.4	1		B

12.13 For $k = 5$, $v = 15$, and $p = 1 - \alpha = 0.95$, $Q_{0.95} = 4.37$ from Appendix S. Since $MSE \approx 8.67$ and $n = 4$, $w \approx 4.37\sqrt{8.67/4} \approx 6.4$. In the summary below, sample means that differ by at most 6.4 are indicated with the same grouping symbol. The population mean associated with the larger of two sample means that do not share a common letter is declared greater than that associated with the smaller of those two sample means. With an overall 95% confidence, we conclude that $\mu_5 < \mu_4$, $\mu_5 < \mu_1$, and the other pairs of means may be equal.

Tukey Grouping		Mean	n	Electrolyte
	A	34	4	5
B	A	40	4	2
B	A	40	4	3
B		41	4	4
B		45	4	1

12.15 Let $\alpha = 0.05$. For $k = 4$, $v = 16$, and $p = 1 - \alpha = 0.95$, $Q_{0.95} = 4.05$. Since $MSE = 0.650$ (obtained in Problem 12.6) and $n = 5$, $w = 4.05\sqrt{0.650/5} \approx 1.46$. Using the following graphical summary, the population mean associated with the larger of two sample means that do not share a common letter is declared greater than that associated with the smaller of those two sample means. With an overall 95% confidence, we conclude that μ_C is the smallest population mean, $\mu_D < \mu_A$, $\mu_D < \mu_B$, and μ_A may equal μ_B.

Tukey Grouping		Mean	n	Supplier
	A	69.0	5	C
B		71.8	5	D
C		73.8	5	A
C		74.2	5	B

12.17 (a) $\bar{y}_1 = {}^{204}\!/_6 = 34.0$; $\bar{y}_2 = {}^{216}\!/_6 = 36.00$; $\bar{y}_3 = {}^{226}\!/_6 \approx 37.7$; $\bar{y}_4 = {}^{199}\!/_6 \approx 33.2$

$\bar{y} = (204 + 216 + 226 + 199)/[(4)(6)] = 845/24 \approx 35.2$

$s^2_{pooled} = (2.00 + 0.80 + 2.27 + 2.17)/4 = 7.24/4 = 1.81$

Since $k = 4$ and $n = 6$, $v = 4(6 - 1) = 20$ and $\sqrt{(k-1)/(kn)} = \sqrt{1/8}$. For $k = 4$, $v = 20$ and $\alpha = 0.01$, $h = 3.42$ from Appendix R. Thus,

$$[LDL, UDL] = \bar{y} \pm h s_{pooled}\sqrt{(k-1)/(kn)}$$

$$\approx 35.2 \pm 3.42\sqrt{1.81}\sqrt{1/8}$$

$$\approx 35.2 \pm 1.6 = [33.6, 36.8].$$

Since $\bar{y}_3$ is greater than UDL (or $\bar{y}_4$ is less than LDL), reject equality of the population means.

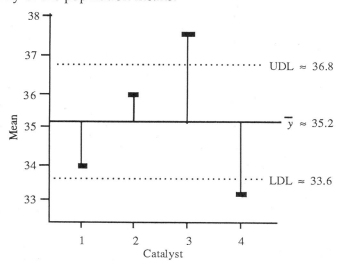

(b) Since $n = 6$ and $k = 4$, SSE has $\nu = nk - k = 24 - 4 = 20$ degrees of freedom. For $k = 4$, $\nu = 20$ and $\alpha = 0.01$, $Q_{0.99} = 5.02$ from Appendix S. Using $MSE = s^2_{pooled} = 1.81$ from (a), $w = 5.02\sqrt{1.81/6} \approx 2.76$.

Tukey Grouping			Mean	n	Catalyst
		A	33.2	6	4
	B	A	34.0	6	1
C	B		36.0	6	2
C			37.7	6	3

In the preceding graphical summary, means that differ by at most $w = 2.76$ share a common letter. A population mean associated with the larger of two sample means that do not share a common letter is declared greater than that associated with the smaller of those two sample means. With an overall 99% confidence, we conclude that $\mu_4 < \mu_2$, $\mu_4 < \mu_3$, and $\mu_1 < \mu_3$.

(c) Using equations (12.5) through (12.7) and the result in (a),

$MSE = s^2_{pooled} = 1.81 = SSE/20$. Thus, $SSE = 20(1.81) = 36.20$. [*Note*: Since $n = 6$ and $k = 4$, SSE has $\nu = k(n - 1) = 4(6 - 1) = 20$ degrees of freedom.]

Using the given totals, $T = 204 + 216 + 226 + 199 = 845$. From equation (12.19),

$$SST = \frac{204^2}{6} + \frac{216^2}{6} + \frac{226^2}{6} + \frac{199^2}{6} - \frac{845^2}{24}$$

$$\approx 29{,}824.8333 - 29{,}751.0417$$

$$= 73.7916 \approx 73.79$$

with $4 - 1 = 3$ degrees of freedom.

From equation (12.15), $SSY = SST + SSE \approx 73.79 + 36.20 = 109.99$. Also, SSY has $nk - 1 = 6(4) - 1 = 23$ degrees of freedom.

Combining the preceding results, we obtain the following ANOVA summary. Since $f = MST/MSE \approx 24.60/1.81 \approx 13.59$ and $F = MST/MSE$ has an F distribution with $\nu_1 = 3$ and $\nu_2 = 20$, the p-value of the F test is $P(F \geq 13.59) \approx 0.000046168 \approx 0.0000$. That value is less than $\alpha = 0.01$, so we reject H_0: "The average yields of the four catalysts are equal" and conclude that the true average yields differ for at least two catalysts.

Source	df	SS	MS	F-ratio	p-value
Catalyst	3	73.79	24.60	13.59	0.0000
Error	20	36.20	1.81		
Total	23	109.99			

12.19

Source	df	SS	MS	F-ratio	p-value
Vendor	2	21.2785	10.6393	38.42	0.0000
Error	10	2.7692	0.2769		
Total	12	24.0477			

Let $T_1 = 7.3 + 8.1 + 8.4 + 7.5 = 31.3$, $T_2 = 10.7 + 10.2 + 10.2 + 10.7 + 9.9 + 11.0$ $= 62.7$, $T_3 = 10.1 + 11.6 + 10.8 = 32.5$, and $T = T_1 + T_2 + T_3 = 126.5$. Also, let $n_1 = 4$, $n_2 = 6$, $n_3 = 3$, and $N = n_1 + n_2 + n_3 = 13$. Using Equation (12.22)

$$SST = \frac{T_1^2}{n_1} + \frac{T_2^2}{n_2} + \frac{T_3^2}{n_3} - \frac{T^2}{N}$$

$$= \frac{31.3^2}{4} + \frac{62.7^2}{6} + \frac{32.5^2}{3} - \frac{126.5^2}{13}$$

$$\approx 1252.2208 - 1230.9423$$

$$= 21.2785 \text{ with 2 degrees of freedom.}$$

Since $SSY = (N-1)s_Y^2 \approx 12(2.003974) \approx 24.0477$ with 12 degrees of freedom, $SSE = SSY - SST \approx 24.0477 - 21.2785 = 2.7692$ with $12 - 2 = 10$ degrees of freedom. Using these results, we obtain the preceding ANOVA table. The p-value is so small that we reject $H_0: \mu_1 = \mu_2 = \mu_3$ and conclude that at least 2 means differ.

For $k = 3$, $v = 10$ and $\alpha = 0.05$, $Q_{0.95} = 3.88$. Using $\bar{y}_1 = 31.3/4 = 7.825$, $\bar{y}_2 = 62.7/6 = 10.450$, $\bar{y}_3 = 32.5/3 \approx 10.833$ and $MSE \approx 0.2769$, the following multiple comparison intervals and conclusions are obtained.

$$\left(\bar{y}_3 - \bar{y}_1\right) \pm w \approx \left(10.833 - 7.825\right) \pm 3.88\sqrt{\frac{0.2769}{2}\left(\frac{1}{3} + \frac{1}{4}\right)}$$

$$\approx 3.008 \pm 1.103 = [1.905, 4.111]; \text{ so, } \mu_3 > \mu_1.$$

$$\left(\bar{y}_3 - \bar{y}_2\right) \pm w \approx \left(10.833 - 10.450\right) \pm 3.88\sqrt{\frac{0.2769}{2}\left(\frac{1}{3} + \frac{1}{6}\right)}$$

$$\approx 0.383 \pm 1.021 = [-0.638, 1.404]; \text{ so,}$$
$$\mu_2 \text{ and } \mu_3 \text{ may be equal.}$$

$$\left(\bar{y}_2 - \bar{y}_1\right) \pm w \approx \left(10.450 - 7.825\right) \pm 3.88\sqrt{\frac{0.2769}{2}\left(\frac{1}{6} + \frac{1}{4}\right)}$$

$$\approx 2.625 \pm 0.932 = [1.693, 3.557]; \text{ so, } \mu_2 > \mu_1.$$

Section 12.1.5

12.21 Let Y_{ij} denote the value for the ith battery in the sample of batteries for the jth electrolyte. Since only 5 electrolytes are under consideration and each is included in the study, this is a fixed-effects situation. A statistical model is

$$Y_{ij} = \mu + \tau_j + \varepsilon_{ij}; \ i = 1,2,3,4; \ j = 1,2,3,4,5; \ \varepsilon_{ij} \sim NID\left(0, \sigma^2\right)$$

where μ denotes the average of $\mu_1, \mu_2, \mu_3, \mu_4$, and μ_5; μ_j denotes the true average value for the jth electrolyte; $\tau_j = \mu_j - \mu$ denotes the true effect of the jth electrolyte; and $\tau_1 + \tau_2 + \tau_3 + \tau_4 + \tau_5 = 0$.

12.23 (a) Since $E[MST] = \sigma^2 + n\phi_\tau$ is given as $\sigma^2 + 7\phi_\tau$ in the ANOVA summary, $n = 7$. Since SSY has $nk - 1$ degrees of freedom, $nk - 1 = 34$ from the ANOVA table. Thus, $k = (34 + 1)/7 = 5$. The fixed effect model is

$$Y_{ij} = \mu + \tau_j + \varepsilon_{ij}; \; i = 1,2,3,4,5,6,7; \; j = 1,2,3,4,5; \; \varepsilon_{ij} \sim NID(0,\sigma^2)$$

where μ denotes the average of $\mu_1, \mu_2, \mu_3, \mu_4$ and μ_5; μ_j denotes the true average value at the jth treatment level; $\tau_j = \mu_j - \mu$ denotes the true effect of the jth treatment level; and $\sum_{j=1}^{5} \tau_j = 0$.

(b) $H_0: \tau_j = 0$ for $j = 1,2,3,4,5$

(c) $H_0: \tau_j = 0$ for $j = 1,2,3,4,5$;

$H_a: \tau_j \neq 0$ for at least one $j \in \{1,2,3,4,5\}$

Test statistic: $F = MST/MSE$; $v_1 = 4$, $v_2 = 30$ (**Note:** Since $k = 5$ and $n = 7$, $v_1 = k - 1 = 4$ and $v_2 = k(n - 1) = 30$.)

Decision rule: Reject H_0 if $f > 2.690$.

Analysis: $SST = SSY - SSE = 514 - 270 = 244$ with 4 degrees of freedom. Thus, $MST = 244/4 = 61$. Since $MSE = 9$, $f = 61/9 \approx 6.778$. Reject H_0.

(d) The p-value is $P(F_{(4,\,30)} \geq f) \approx P(F_{(4,\,30)} \geq 6.778) \approx 0.00052$. Therefore, H_0 can be rejected for any $\alpha \geq 0.00052$.

Section 12.1.6

12.25

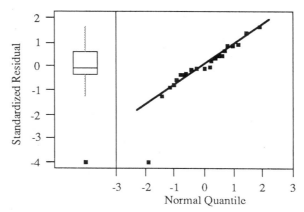

Consider the preceding JMP box plot and normal quantile plot. The indicated outlier is associated with the value of 50 obtained for furnace D. If possible, the reason for obtaining such a value should be determined. If the unusually small observation is due to a special cause, further analyses could be conducted after removing that observation.

Assuming that a special cause for the outlier has been identified and recalculating, the following normal quantile plot is obtained. The Shapiro-Wilk W test and the linear pattern in the plotted points support normality.

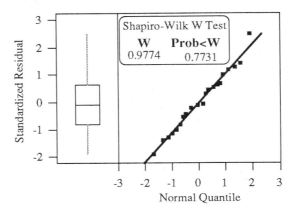

A plot of the standardized residuals versus furnaces follows. The standardized residual for the furnace 2 reading of 95 (indicated by the heavier dot on the scatter plot) exceeds 2. That residual lies just off the line in the preceding normal quantile plot. No outliers (mild or strong) are indicated on the preceding box plot, so we will accept the residual as an observation from the upper tail of a normal distribution. The points are reasonably symmetric about the dotted horizontal line through 0, so the model seems to provide a reasonable fit.

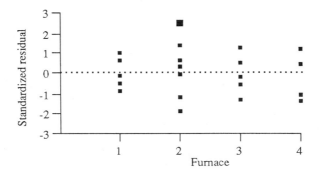

The following JMP range chart for the standardized residuals supports an assumption of equal variances. Since the scatter plot does not provide strong evidence against that assumption, we will assume that the population variances are approximately equal.

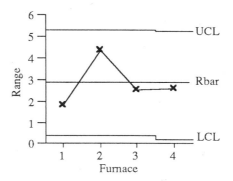

(**Note:** The preceding range chart was calculated for unequal sample sizes, since one observation was removed from the furnace 4 sample. The JMP module modified the range chart accordingly.)

12.27 The following Shapiro-Wilk W test for normality (p-value = 0.0000) and normal quantile plot of the standardized residuals suggest non-normality. The accompanying box plot reveals many outliers.

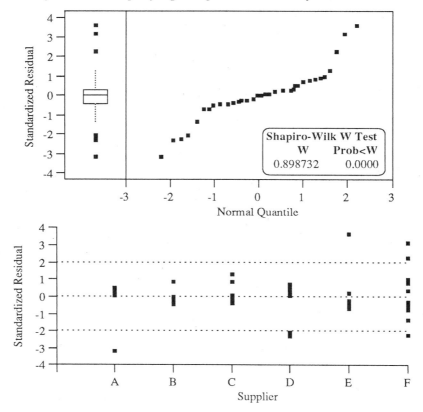

The preceding plot of standardized residual versus supplier shows that only the samples for suppliers B and C have no extreme standardized residuals. For the other suppliers, at least one standardized residual less than -2 or greater than 2 is present.

The preceding scatter plot also seems to indicate that variances differ. The following R chart does not support that assumption. However, sample sizes exceed 10, so an s chart may be more appropriate.

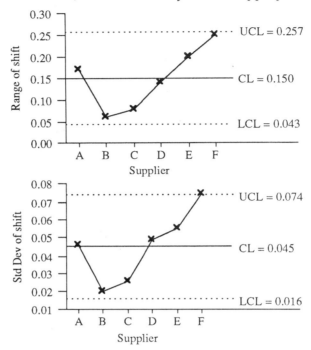

When an s chart is prepared, the standard deviation of the supplier F data exceeds the upper control limit. This indicates that an assumption of equal variances should not be made.

[**Note:** The patterns in the R and s charts should be ignored. The naming and ordering of the suppliers was arbitrary.]

Section 12.1.7

12.31 (a) H_0: "The 5 distributions of abrasive wear are identical."
H_a: "At least two of the 5 population medians differ."

Test statistic: $H_{KW} = \frac{1}{35} \times \left[\frac{T_1^2}{4} + \frac{T_2^2}{4} + \frac{T_3^2}{4} + \frac{T_4^2}{4} + \frac{T_5^2}{4} \right] - 63$

[**Note:** Since 4 observations were obtained for each of the 5 materials, a total of $N = 20$ observations were considered. Thus, $12/[N(N + 1)] = 12/420 = 1/35$ and $3(N + 1) = 63$.]

Decision rule: Let $\alpha = 0.05$. Assuming H_{KW} has an approximate chi-square distribution with 4 degrees of freedom, reject H_0 if $h_{KW} \geq 9.488$.

(**Note:** The sample sizes are less than the recommended value of at least 5. Consulting Table X.8 in *Basic Statistical Tables* (1971), published by The Chemical Rubber Company, $P(H_{KW} \geq 5.6923) \approx 0.049$. Thus, our decision rule is quite conservative.)

(b) Consider the following summary of ranks and rank totals. Since

$$h_{KW} = \frac{1}{35} \times \left(\frac{21.0^2 + 69.0^2 + 42.0^2 + 63.0^2 + 15.0^2}{4} \right) - 63 \approx 16.714 \text{ exceeds } 9.488,$$

reject H_0 and conclude that at least two population medians differ. If the distributions of abrasive wear are symmetric, the test is for equality of means and the conclusion is the same as that in Table 12.7.

Material									
1		2		3		4		5	
y	Rank	y	Rank	y	Rank	y	Rank	y	Rank
23	2	42	14.5	37	9	41	13	20	1
25	4.5	44	16.5	38	10	42	14.5	25	4.5
30	6.5	45	18	39	11.5	44	16.5	24	3
31	8	50	20	39	11.5	49	19	30	6.5
	21.0		69.0		42.0		63.0		15.0

(c) To have an overall multiple comparison level of 5%, each of the ten rank sum tests must be conducted at the 0.5% level. Since significance levels of that magnitude are not provided in Appendix P, a more extensive table or a computer program must be used. Most computer programs include the equivalent Mann-Whitney U test as an alternative to the Wilcoxon rank sum test.

Selecting **Stat/Nonparametrics/Mann-Whitney** in Minitab gives a confidence interval for the difference in the two population medians and a normal approximation to the p-value for the Mann-Whitney U test. A continuity correction is used with the normal approximation and, when appropriate, an adjustment is made for ties. A Minitab summary for material 1 versus material 2 follows.

```
Mann-Whitney Confidence Interval and Test
MTRL_1        N = 4     Median =      27.500
MTRL_2        N = 4     Median =      44.500
Point estimate for ETA1–ETA2 is          -19.000
97.0 Percent C.I. for ETA1–ETA2 is (-26.998, -11.000)
W = 10.0
Test of ETA1 = ETA2   vs.   ETA1 ~= is significant at 0.0304
```

167

Minitab uses η (the lowercase Greek "eta") to denote a population median. In the preceding summary, the p-value (Test of ETA1 = ETA2 vs. ETA1 ~= ETA2 is significant at 0.0304) exceeds 0.005, so we cannot reject H_0: $M_1 = M_2$ in favor of H_a: $M_1 \neq M_2$.

The Minitab p-values for the 10 comparisons are summarized in the following table. For each test, the p-value exceeds 0.005. We are unable to reject equality of population medians for any of the 10 pairs.

Comparison	p-value	Comparison	p-value
1 versus 2	0.0304	2 versus 4	0.4651*
1 versus 3	0.0304	2 versus 5	0.0304
1 versus 4	0.0304	3 versus 4	0.0294*
1 versus 5	0.4651*	3 versus 5	0.0304
2 versus 3	0.0294*	4 versus 5	0.0304

*Adjusted for ties

In Example 12.10, Tukey's procedure enabled us to conclude that the true average wears of materials 1 and 5 are significantly less than those of materials 2, 3, and 4. The small sample sizes and very conservative Bonferroni procedure prevent us from making similar conclusions about the population medians.

Section 12.2

12.37 (a) $Y_{ij} = \mu + \beta_i + \tau_j + \varepsilon_{ij}$, where

$\beta_i = \mu_{i\bullet} - \mu$ denotes the true effect of the ith day; $i = 1, 2, 3$;

$\tau_j = \mu_{\bullet j} - \mu$ denotes the true effect of the jth filter type; $j = 1, 2, 3, 4, 5$;

$\varepsilon_{ij} \sim \text{NID}(0,\sigma^2)$; $\sum_{i=1}^{3} \beta_i = 0$; and $\sum_{j=1}^{5} \tau_j = 0$

H_0: $\tau_j = 0$ for $j = 1,2,3,4,5$

H_a: $\tau_j \neq 0$ for at least one $j \in \{1,2,3,4,5\}$

Test statistic: $F = MST/MSE$; $v_1 = 4$, $v_2 = 8$
[**Note:** Since $k = 5$ and $b = 3$, $v_1 = k - 1 = 4$ and $v_2 = (k - 1)(b - 1) = 8$.]

Decision rule: Reject H_0 if $P(F_{(4,\ 8)} \geq f) \leq \alpha$.

(b)

Analysis of Variance (Balanced Designs)

Factor	Type	Levels	Values
Day	fixed	3	1 2 3
Type	fixed	5	1 2 3 4 5

Analysis of Variance for Rdg

Source	DF	SS	MS	F	P
Day	2	1.6480	0.8240	0.94	0.429
Type	4	11.5560	2.8890	3.31	0.071
Error	8	6.9920	0.8740		
Total	14	20.1960			

Selecting **Stat/ANOVA/Balanced ANOVA...** in Minitab and specifying the tabled reading as the response, with day and type of filter as factors, produces the following summary. Since days were used as blocks, we ignore the included test on day. The p-value of the test on filter type is 0.071, so $H_0: \tau_j = 0$ $(j = 1, 2, 3, 4, 5)$ can be rejected for any $\alpha \geq 0.071$.

(c) For $k = 5$, $\nu = 8$, and $\alpha = 0.05$, $Q_{0.95} = 4.89$. Since MSE = 0.8740 in (b) and $n = 3$, $w = 4.89\sqrt{0.8740/3} \approx 2.64$. The population means associated with two sample means that differ by at least 2.64 can be declared significantly different. The largest of the paired differences is $\bar{x}_5 - \bar{x}_4 = 2.50$. So, any two sample means differ by at most 2.50. Using an experimentwise error rate of 0.05, no two population means can be declared significantly different.

A graphic for the preceding discussion follows. Since all filter types share a common letter, no two population means can be declared significantly different.

Tukey Grouping	Mean	n	Filter type
A	16.30	3	4
A	16.97	3	1
A	17.47	3	3
A	18.17	3	2
A	18.80	3	5

(d) The following normal quantile plot for the standardized residuals is reasonably linear and the box plot reveals no outliers. These observations and the Shapiro-Wilk W test (p-value = 0.9011) support the normality assumption.

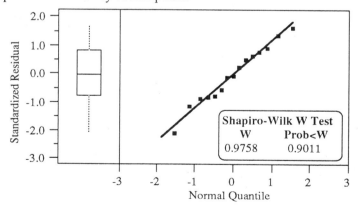

The following plot of the standardized residuals versus days seems to indicate that variability increases with time. The standardized residual that is less than -2.0 causes most of that impression. That residual is associated with the reading

169

obtained on the 3rd day for the type 2 filter. The conditions at the time that reading was obtained should be considered to see if a special cause was present.

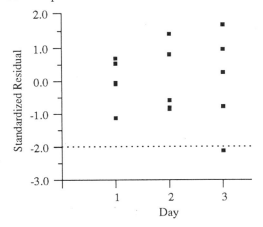

Consider the following plot of the standardized residual versus type of filter. The residual below the reference line at -2 also plotted below the reference line on the preceding plot of standardized residual versus day.

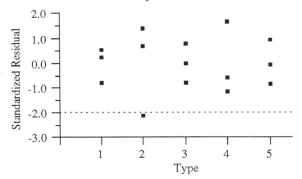

12.39 (a) $Y_{ij} = \mu + \beta_i + \tau_j + \varepsilon_{ij}$, where
$\beta_i = \mu_{i\bullet} - \mu$ denotes the true effect of the ith sample; $i = 1, 2, \ldots , 10$;
$\tau_j = \mu_{\bullet j} - \mu$ denotes the true effect of the jth laboratory; $j = 1, 2$;
$\varepsilon_{ij} \sim \text{NID}(0,\sigma^2)$; $\sum_{i=1}^{10}\beta_i = 0$; and $\sum_{j=1}^{2}\tau_j = 0$.

$H_0: \tau_1 = \tau_2; H_a: \tau_1 \neq \tau_2$

Test statistic: $F = MST/MSE$; $v_1 = 1, v_2 = 9$
[**Note:** Since $k = 2$ and $b = 10$, $v_1 = k - 1 = 1$ and $v_2 = (k - 1)(b - 1) = 9$.]

Decision rule: Reject H_0 if $P(F_{(1, 9)} \geq f) \leq \alpha$.

170

(b) Selecting **Stat/ANOVA/Balanced ANOVA...** in Minitab and specifying the tabled reading as the percent, with sample and lab as factors, produces the following ANOVA summary. Since the split samples were used as blocks, we ignore the test on sample. The p-value of the test on laboratories is 0.153, so we have insufficient evidence to say that there is a true laboratory effect.

Analysis of Variance for Percent

Source	DF	SS	MS	F	P
Sample	9	2.17648	0.24183	34.95	0.000
Lab	1	0.01682	0.01682	2.43	0.153
Error	9	0.06228	0.00692		
Total	19	2.25558			

(c) In the following JMP normal quantile plot, the standardized residuals associated with sample 1 differ enough from the other standardized residuals that those readings should be reconsidered (2.49 for the supplier lab; -2.49 for the plant lab). No evidence against normality is obtained from the Shapiro-Wilk W test result (p-value = 0.6093).

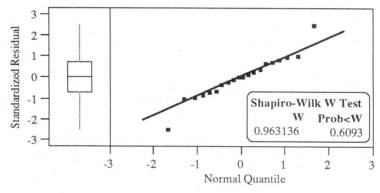

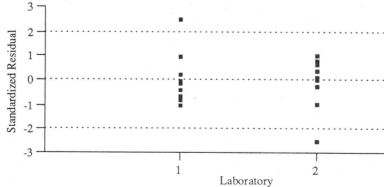

The preceding scatter plot of the standardized residual versus laboratory supports an assumption of equal variances for the laboratories.

(d) The following scatter plot of the 10 sample pairs gives the impression that the supplier laboratory tends to report lower ash content than that reported by the laboratory at the steel plant. The supplier reading is above the line $y = x$ for only 2 of the 10 pairs.

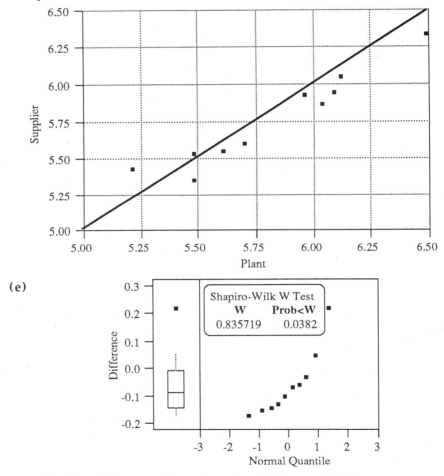

(e)

Letting diff = supplier - plant, the 10 observed differences are 0.22, -0.13, 0.05, -0.06, -0.03, -0.14, -0.15, -0.07, -0.17, and -0.10. Using the **Analyze** menu in JMP, the preceding normal quantile plot, box plot, and Shapiro-Wilk W test result are obtained. The curvature in the normal quantile plot, small p-value (0.0382) of the Shapiro-Wilk W test, and outlier indicated by the box plot suggest that normality should not be assumed for the population of differences. This may be due to the outlier, which is associated with sample 1. If further investigation indicates that the sample 1 data are valid, these results indicate that the differences are from a skewed distribution. In such cases, the t procedure should not be used. If the outlier is present because a special cause was present (e.g., incorrect reading or recording error), further analyses can be conducted after removing that outlier.

Assuming the 10 data values are valid, the one-sample sign test (Section 7.7.1) can be used to test $H_0: M_D = 0$ versus $H_a: M_D \neq 0$. Since $r = 2$ observed differences are positive and $P(R \leq 2) = 0.0546875$ when $R \sim b(10, 0.5)$, the p-value of this test is $2(0.0546875) = 0.109375 \approx 0.1094$. There is insufficient evidence to conclude that the median of the population of differences is not 0.

Section 12.2.6

12.41 (a) H_0: "The 3 environment populations have identical distributions."
H_a: "At least 2 populations have different medians."

Test Statistic: $F_r = 12(SSK)/[3(3 + 1)] = SSK$ with SSK the sum of squares of the treatment ranks when the treatment observations are ranked *within blocks*.

Decision Rule: Reject H_0 if $P(F_r > f_r) < \alpha$.

$$\text{Use } P(F_r > f_r) \approx P(\chi^2_{(2)} > f_r).$$

Blocks (Bars)	Treatments (Environments)			Total
	A	B	C	
1	1	2	3	6
2	1	2	3	6
3	1	2	3	6
4	1	2	3	6
5	1	2	3	6
6	1	2	3	6
Total	6	12	18	36

Analysis: Ranks of the treatment observations within blocks are summarized in the preceding table. Since

$$SSK = \frac{6^2}{6} + \frac{12^2}{6} + \frac{18^2}{6} - \frac{36^2}{18} = 12,$$

$f_r = 12$. For $W \sim \chi^2_{(2)}$, $P(F_r > 12) \approx P(W > 12) \approx 0.0025$.

Reject H_0 and conclude that at least one of $M_A \neq M_B$, $M_A \neq M_C$, or $M_B \neq M_C$ is true.

(b) Since $C(3, 2) = 3$ comparisons must be made at a multiple comparison rate of 0.03, each test should be conducted at the $0.03/3 = 0.01$ significance level. Appendix P does not contain critical values for a two-sided test with $\alpha = 0.01$. Thus, a computer program with the Wilcoxon Rank Sum procedure, or the equivalent Mann-Whitney U procedure, should be used.

SYSTAT calculates the Mann-Whitney U statistic and p-value when **Stats/Npar/Kruskal-Wallis...** is selected and only 2 treatment levels are used. For environments A and B, the following summary is obtained. Since the p-value (0.064) exceeds 0.01, we will not reject $H_0: M_A = M_B$. [**Note:** Environments A, B, and C are denoted 1, 2, and 3, respectively, in the summaries.]

KRUSKAL-WALLIS ONE-WAY ANALYSIS OF VARIANCE FOR 12 CASES
DEPENDENT VARIABLE IS READING
GROUPING VARIABLE IS ENVIRON

GROUP	COUNT	RANK SUM
1	6	27.5
2	6	50.5

MANN-WHITNEY U TEST STATISTIC = 6.500
PROBABILITY IS 0.064
CHI-SQUARE APPROXIMATION = 3.427 WITH 1 DF

Summaries for the other two comparison tests follow. Since both p-values are less than 0.01, conclude that $M_A \neq M_C$ and $M_B \neq M_C$.

KRUSKAL-WALLIS ONE-WAY ANALYSIS OF VARIANCE FOR 12 CASES
DEPENDENT VARIABLE IS READING
GROUPING VARIABLE IS ENVIRON

GROUP	COUNT	RANK SUM
1	6	21
3	6	57

MANN-WHITNEY U TEST STATISTIC = 0.000
PROBABILITY IS 0.004
CHI-SQUARE APPROXIMATION = 8.308 WITH 1 DF

KRUSKAL-WALLIS ONE-WAY ANALYSIS OF VARIANCE FOR 12 CASES
DEPENDENT VARIABLE IS READING
GROUPING VARIABLE IS ENVIRON

GROUP	COUNT	RANK SUM
2	6	22
3	6	56

MANN-WHITNEY U TEST STATISTIC = 1.000
PROBABILITY IS 0.006
CHI-SQUARE APPROXIMATION = 7.462 WITH 1 DF

12.45 **(a)** Selecting **Stat/Nonparametrics/Friedman...** in Minitab and specifying the metal thickness as the response, position as the treatment, and wafer as the block produces the following summary. Using $f_r \approx 45.85$ from that summary (denoted S and not adjusted for ties) with the formula in Problem 12.43, we obtain $f = 18(45.85)/[19(4) - 45.85] \approx 27.37$. The approximate p-value for the test is $P(F_{(4, 72)} \geq 27.37) \approx 0.0000$. Reject H_0 and conclude that the true median thicknesses differ for at least 2 positions.

```
Friedman test of Thcknss by Position blocked by Wafer
S = 45.85  d.f. = 4  p = 0.000
S = 54.97  d.f. = 4  p = 0.000 (adjusted for ties)
```

Position	N	Est. Median	Sum of RANKS
1	19	16.000	73.5
2	19	16.000	76.0
3	19	15.910	41.5
4	19	16.000	70.5
5	19	15.830	23.5

Grand median = 15.948

[**Note:** The advantage of the F approximation is that it can be obtained directly by conducting an analysis of variance on the ranks. The F ratio is automatically adjusted when there are ties.]

Selecting **Stat/ANOVA/Balanced ANOVA...** , after preparing a Minitab worksheet that includes the ranks within blocks, and specifying Rank as the response with Wafer and Position as factors produces the following ANOVA summary. The F ratio in that summary is greater than that obtained from the formula because of the large number of ties in the data set.

Analysis of Variance for Rank

Source	DF	SS	MS	F	P
Wafer	18	0.0000	0.0000	0.00	1.000
Position	4	114.6316	28.6579	47.04	0.000
Error	72	43.8684	0.6093		
Total	94	158.5000			

(b) To have an overall multiple comparison level of 10%, each of the ten rank sum tests must be conducted at the 1.0% level. Significance levels of that magnitude are not provided in Appendix P, so a more extensive table or a computer program must be used. Most computer programs include the equivalent Mann-Whitney U test as an alternative to the Wilcoxon rank sum test.

Selecting **Stat/Nonparametrics/Mann-Whitney** in Minitab gives a confidence interval for the difference in the two population medians and a normal approximation to the p-value for the

175

Mann-Whitney U test. A continuity correction is used with the normal approximation, and an adjustment is made for ties (when appropriate). Minitab uses η (the lowercase Greek "eta") to denote a population median.

```
Mann-Whitney  Confidence  Interval  and  Test
Pstn_1         N = 19     Median =        16.080
Pstn_2         N = 19     Median =        16.000
Point estimate for ETA1–ETA2 is           0.000
99.1 Percent C.I. for ETA1–ETA2 is  (-0.090, 0.080)
W  =  373.0
Test of ETA1 = ETA2   vs.   ETA1 ~= is significant at 0.9534
The test is significant at 0.9505 (adjusted for ties)
Cannot reject at alpha = 0.05
```

In the preceding summary for position 1 versus position 2, the p-value (0.9505 when adjusted for ties) exceeds 0.01, so we cannot reject $H_0: M_1 = M_2$ in favor of $H_a: M_1 \neq M_2$.

The Minitab p-values for the 10 comparisons are summarized in the following table. At a multiple comparison rate of 10%, we conclude that $M_1 \neq M_3$, $M_1 \neq M_5$, $M_2 \neq M_3$, $M_2 \neq M_5$, $M_3 \neq M_4$, $M_3 \neq M_5$, and $M_4 \neq M_5$.

Comparison	p-value*	Comparison	p-value*
1 versus 2	0.9505	2 versus 4	0.4638
1 versus 3	0.0025	2 versus 5	0.0000
1 versus 4	0.5331	3 versus 4	0.0033
1 versus 5	0.0000	3 versus 5	0.0050
2 versus 3	0.0003	4 versus 5	0.0000

*Adjusted for ties

Chapter 12 Supplementary Problems

12.47 (a) The following grouped box plot, normal quantile plots and Shapiro-Wilk W test results were obtained using JMP Version 3.1.5. The normal quantile plots include the Lillefors graph as a test for normality. If any points fall outside the Lillefors bounds (indicated by the dotted curves), the sampled population should be considered nonnormal. Otherwise, the sample could have come from a normally distributed population. The Lillefors graph for sample 1 and the small p-value (0.0062) of the Shapiro-Wilk W test provide strong evidence that the population of forces for shift 1 is not normally distributed. The distributions for the other two shifts may be normally distributed.

JMP includes means diamonds with a box plot. The vertical tips of a diamond mark the endpoints of a 95% confidence interval for the mean, the horizontal tips of a diamond mark the location of the sample mean, and the solid segment within the box marks the location of the sample median. Since the means diamond for shift 1 does not overlap the other two means diamonds, we have reason to believe that the average force is least for shift 1.

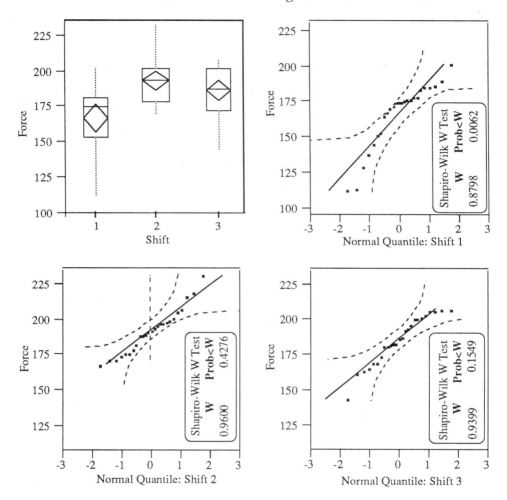

(b) The following grouped box plot indicates an outlier (1.38) in the modified shift 1 data. An investigation should follow to see if that is due to the presence of a special cause.

All points on the normal quantile plot of the modified shift 1 data are within the Lillefors bounds. The p-value of the Shapiro-Wilk W tests is not small. There is no strong evidence against an assumption of normality. [**Note:** Only the shift 1 data were modified. Normal quantile plots of the data for shifts 2 and 3 are included in (a).]

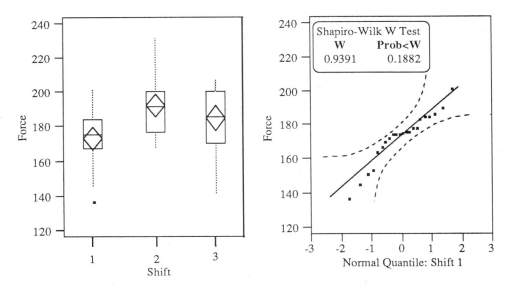

(c) Consider the following JMP ANOVA summaries. The p-value of the F test is so small (0.0004) that we reject $H_0: \mu_1 = \mu_2 = \mu_3$ and conclude that at least two population means differ.

Analysis of Variance

Source	DF	Sum of Squares	Mean Square	F Ratio
Model	2	4616.852	2308.43	8.8731
Error	69	17951.135	260.16	**Prob>F**
C Total	71	22567.986		0.0004

Means for Oneway Anova

Level	Number	Mean	Std Error
1	22	173.455	3.4388
2	25	193.120	3.2259
3	25	186.280	3.2259

Std Error uses a pooled estimate of error variance

For $k = 3$, $\nu = 69$, and $\alpha = 0.05$, $Q_{0.95} \approx 3.31$ from Appendix S; $MSE = 260.16$ from the preceding ANOVA summary; and

$$w = 3.31\sqrt{(260.16/2)[(1/n_i) + (1/n_j)]} = 3.31\sqrt{130.08[(1/n_i) + (1/n_j)]}.$$

Using this information and the means in the preceding summary, we obtain the following multiple comparison intervals and conclusions.

$$\left(\bar{y}_2 - \bar{y}_1\right) \pm w = (193.120 - 173.455) \pm 3.31\sqrt{130.08\left(\tfrac{1}{25} + \tfrac{1}{22}\right)}$$

$$\approx 19.665 \pm 11.036 = [8.629, 30.701]; \text{ so, } \mu_2 > \mu_1.$$

178

$$\left(\bar{y}_2 - \bar{y}_3\right) \pm w = (193.120 - 186.280) \pm 3.31\sqrt{130.08\left(\tfrac{1}{25} + \tfrac{1}{25}\right)}$$
$$\approx 6.840 \pm 10.678 = [-3.838, 17.518]; \text{ so, } \mu_2 \text{ and } \mu_3$$
may be equal.

$$\left(\bar{y}_3 - \bar{y}_1\right) \pm w = (186.280 - 173.455) \pm 3.31\sqrt{130.08\left(\tfrac{1}{25} + \tfrac{1}{22}\right)}$$
$$\approx 12.825 \pm 11.036 = [1.789, 23.861]; \text{ so, } \mu_3 > \mu_1.$$

(d) $Y_{ij} = \mu + \tau_j + \varepsilon_{ij} \, ; j = 1, 2, 3; i = 1, 2, \ldots, n_j; n_1 = 22, n_2 = n_3 = 25;$
$\varepsilon_{ij} \sim \text{NID}(0,\sigma^2); \mu$ denotes the average of the μ_j; and $\tau_j = \mu_j - \mu.$

The following histogram for the standardized residuals is reasonably bell-shaped and the box plot is reasonably symmetric about the median. The points on the normal quantile plot lie within the Lillifors bounds and the p-value of the Shapiro-Wilk W test (0.8287) is large. These observations indicate that a normality assumption is reasonable.

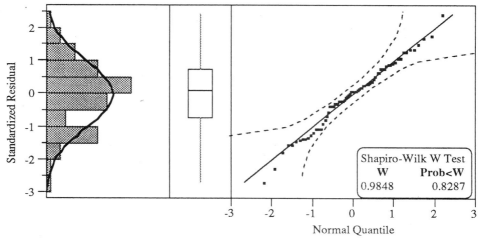

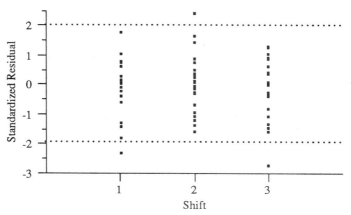

179

The preceding plot of the standardized residual versus shift supports an assumption of equal variances. Notice however, that 3 standardized residuals fall outside the interval [-2, 2]. Those are associated with 138, 232, and 144 from shifts 1, 2, and 3, respectively. The conditions during the times when the associated subassemblies were manufactured and tested should be investigated thoroughly.

Considerations to this point indicate that the model assumptions are adequately satisfied. However, the model accounts for very little of the variability in force. The coefficient of determination is $R^2 = SST/SSY \approx 4616.852/22{,}567.986 \approx 0.204575$, so the model accounts for approximately 20.5% of the variability in force. To account for more of the variability, additional factors must be included in the model.

(e) The following histogram for the modified set of 72 observations is reasonably bell-shaped. The points on the normal quantile plot fall along a line and lie within the Lillefors bounds. The p-value (0.8892) of the Shapiro-Wilk W test for normality is large. An assumption of normality seems reasonable.

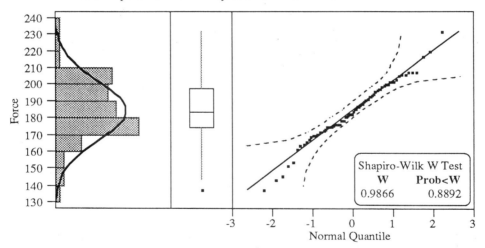

One outlier (138, from Shift 1) is indicated on the preceding box plot. That outlier was indicated on the grouped box plots in (b). In this context, a value of 138 looks as if it could be from the lower tail of a normal distribution.

(f) Adding the lower specification limit to the histogram for the modified set of 72 measurements, we see that very few subassemblies will fail to meet a lower specification limit of 134 Newtons. From the following JMP summary, the sample mean and

180

standard deviation (to the nearest 10th) are 184.7 N and 17.8 N, respectively. If the force (denoted Y) is normally distributed with $\mu = 184.7$ and $\sigma = 17.8$, $P[Y < 134] = P[Z < (134 - 184.7)/17.8] \approx \Phi(-2.85) = 0.00219$. So, about 0.22% of the day's production of subassemblies failed to meet specifications.

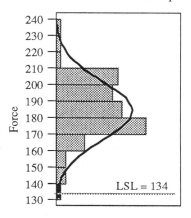

Moments	
Mean	184.7
Std Dev	17.8
Std Error Mean	2.1
Upper 95% Mean	188.9
Lower 95% Mean	180.5
N	72.0
Sum Weights	72.0
Sum	13301
Variance	317.9
Skewness	-0.1
Kurtosis	0.3
CV	9.7

At this point, a careful consideration of the conditions at the time the Shift 1 data were obtained is in order. Questions to consider include:

- "Was there a problem with one of the welding machines?"
- "Was an adequate measurement process used with the Shift 1 data?"
- "Were the Shift 1 lightpipes from a different lot or supplier than those used by the other shifts?".

(g) If shift is not considered, we fail to notice that the average force for Shift 1 may be less than those for the other shifts.

181

CHAPTER 13

DESIGN AND ANALYSIS OF MULTIFACTOR EXPERIMENTS

Section 13.1.1

13.1 **(a)** Let $Y = \beta_0 + \beta_1 x_1 + \beta_2 x_2 + \beta_3 x_1 x_2 + \varepsilon$, where x_1 and x_2 denote atmosphere and back zone temperature, respectively,

$$x_1 = \begin{cases} -1 \text{ when the oxidizing atmosphere is used} \\ +1 \text{ when the reducing atmosphere is used} \end{cases}, \text{ and}$$

$$x_2 = \begin{cases} -1 \text{ when the back zone temperature is 2090 degrees} \\ +1 \text{ when the back zone temperature is 2110 degrees} \end{cases}.$$

(b)

Term	Estimate	Std Error	t Ratio	Prob $>|t|$
Intercept	8.0875	0.1317	61.41	0.0000
Atmosphere	0.1375	0.1317	1.04	0.3554
Temperature	-0.9875	0.1317	-7.50	0.0017
Interaction	-1.2875	0.1317	-9.78	0.0006

The preceding JMP summary reveals a significant interaction. This indicates that a comparison of the four treatment combination means is appropriate.

The following profile plot of the sample means seems to indicate that the treatment effect is least when the reducing atmosphere is used with a back zone temperature of 2110 degrees. Before making any recommendations, we will see if Tukey's procedure supports this observation.

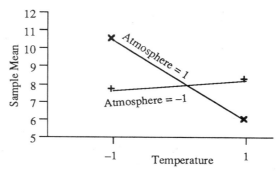

To conduct Tukey's multiple comparison procedure, we first obtain the following ANOVA summary for the model $Y_{ijk} = \mu + \tau_{ij} + \varepsilon_{ijk}$; $i = 1, 2; j = 1, 2; k = 1, 2;$ where Y_{ijk} denotes the kth value for the ith

182

atmosphere and jth temperature combination, τ_{ij} denotes the true effect of the ith atmosphere and jth temperature treatment level, and ε_{ijk} denotes the deviation of the kth value at the same treatment level from the population mean at that treatment level.

Analysis of Variance				
Source	DF	Sum of Squares	Mean Square	F Ratio
Model	3	21.213750	7.07125	50.9640
Error	4	0.555000	0.13875	**Prob>F**
C Total	7	21.768750		0.0012

From the preceding ANOVA summary, $MSE = 0.13875$ and $\nu = 4$ degrees of freedom are associated with error. Letting $k = 4$, $\nu = 4$, and $\alpha = 0.05$, $Q_{0.95} = 5.76$ from Appendix S. Since $n = 2$ observations were obtained at each treatment combination, $w = 5.76\sqrt{0.13875/2} \approx 1.52$. If two sample means differ by at least 1.52 units, conclude that the population mean associated with the larger sample mean is the greater of the two population means.

The observed means for the treatment combinations $1 = (-1, -1)$, $2 = (-1, 1)$, $3 = (1, -1)$, and $4 = (1, 1)$ are 7.65, 8.25, 10.50, and 5.95, respectively. Placing those means in ascending order and comparing the differences to 1.52 gives the following graphic for Tukey's multiple comparison procedure. With a 5% experiment-wise error, we conclude that μ_4 is the least of the four population means, μ_3 is the greatest, and μ_1 may equal μ_2. Since smaller is better, the reducing atmosphere at a back zone temperature of 2110 degrees should be used. The reducing atmosphere at a back zone temperature of 2090 degrees should be avoided.

Population Label and Sample Mean

$4 = (1, 1)$	$1 = (-1, -1)$	$2 = (-1, 1)$	$3 = (1, -1)$
5.95	7.65	8.25	10.50

(c)

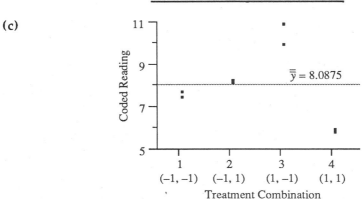

183

The preceding scatter plot of the observations at the treatment levels gives some indication that the variability at the 3rd treatment level is greater than that at the other levels. The 4 sample ranges are 0.3, 0.1, 1.0, and 0.1. $D_4 = 3.267$ for samples of size $n = 2$, and the average of the 4 sample ranges is 0.375. Thus, the upper control limit for the sample range is $(3.267)(0.375) \approx$ 1.225. The 4 ranges are less than 1.225, so the assumption of equal variances may be adequately satisfied.

A coefficient of determination of $R^2 = 0.974505$ was included in the JMP output (not shown). This indicates that our model accounts for approximately 97% of the variability in the data. Also, the points in the following plot of y versus

$$\hat{y} = 8.0875 + 0.1375x_1 - 0.9875x_2 - 1.2875x_1x_2$$

follows the line $y = \hat{y}$ (indicated by the dashed diagonal line in the following graphic). The fit of this model to the data in the sample is a good one.

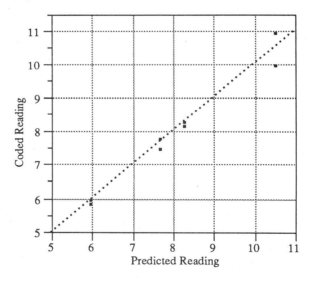

The following normal quantile plot and Shapiro-Wilk W test were obtained using JMP Version 3.1.5. The normal quantile plot includes the Lillefors graph as a test for normality. If any points fall outside the Lillefors bounds (indicated by the dotted curves), the sampled population should be considered nonnormal. Other-wise, the sample could have come from a normally distributed population. In this case, the plot supports an assumption of normality. The Shapiro-Wilk W test (p-value = 0.6222) also supports that assumption.

184

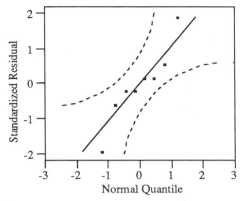

Section 13.1.4

13.3

Treatment	Linear Contrast Coefficients			
Combination	A	B	AB	Total
(1)	−1	−1	+1	T_1
a	+1	−1	−1	T_2
b	−1	+1	−1	T_3
ab	+1	+1	+1	T_4

$$C_A = -T_1 + T_2 - T_3 + T_4 \; ; \; C_B = -T_1 - T_2 + T_3 + T_4 \; ; \; C_{AB} = T_1 - T_2 - T_3 + T_4$$

13.5 **(a)** The following orthogonal table was obtained for A = screw position, B = hold pressure, and C = velocity. From that table, it appears that screw position has the most effect on shrink.

Treatment	Linear Contrast Coefficients							
Combination	A	B	C	AB	AC	BC	ABC	Total
(1)	-1	-1	-1	+1	+1	+1	-1	1.84
a	+1	-1	-1	-1	-1	+1	+1	0.96
b	-1	+1	-1	-1	+1	-1	+1	1.87
ab	+1	+1	-1	+1	-1	-1	-1	0.71
c	-1	-1	+1	+1	-1	-1	+1	1.82
ac	+1	-1	+1	-1	+1	-1	-1	1.36
bc	-1	+1	+1	-1	-1	+1	-1	1.94
abc	+1	+1	+1	+1	+1	+1	+1	0.70
Estimated Effect	-3.74	-0.76	0.44	-1.06	0.34	-0.32	-0.50	11.20

(b)

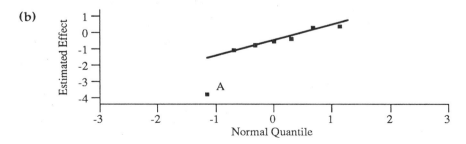

185

(c)

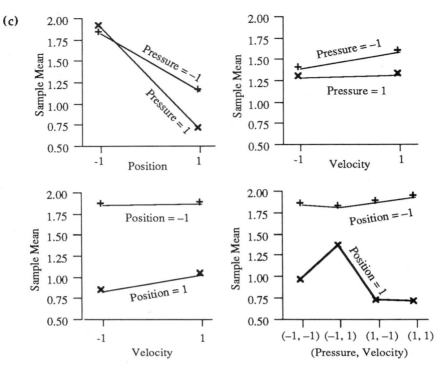

(d) The point associated with A (screw position) lies well off a line through the main body of points in (b). Thus, the effect of A seems to be significant. In (c), the shrink is least at the high screw position. It appears that the higher screw position should be used.

Section 13.1.5

13.9 Let A and B denote "machine design" and "country," respectively, with "standard" the low level of A and "domestic" the low level of B. We first test for an interaction effect. If a significant result is obtained, the means at $(-1, -1)$, $(-1, 1)$, $(1, -1)$, and $(1, 1)$ will be compared. Otherwise, the effects of A and B will be considered individually.

$H_0: \Psi_{AB} = 0; H_a: \Psi_{AB} \neq 0$

Test statistic: $T = \dfrac{C_{AB}}{\sqrt{20(MSE)}}$, with $\nu = 16$

Analysis: From the following table, $C_{AB} = -312$ and $MSE = 1183.6/4 =$ 295.9. Thus, $t = -312/\sqrt{20(295.9)} \approx -4.06$. Since the p-value is $2P(T > |-4.06|) \approx 0.0009$, we can reject H_0 for any reasonable value of α.

186

[**Note:** The same result was obtained in Problem 13.2, using a multiple regression analysis. See the solution of that problem for further analyses.]

Treatment Combination	Linear Contrast Coefficients			Total	Sample Variance
	A	B	AB		
(1)	−1	−1	+1	1003	145.3
a	+1	−1	−1	1213	249.8
b	−1	+1	−1	1173	598.8
ab	+1	+1	+1	1071	189.7
Contrast	108	28	−312	4460	1183.6

13.11 **(a)** Let A, B, and C denote "power", "SF6 flow rate", and "oxygen flow rate," respectively. The following orthogonal table will be used to analyze the effects of these factors and their interactions.

tc	Linear Contrast Coefficients							Total	s^2
	A	B	C	AB	AC	BC	ABC		
(1)	-1	-1	-1	+1	+1	+1	-1	8.45	0.040292
a	+1	-1	-1	-1	-1	+1	+1	10.11	0.000492
b	-1	+1	-1	-1	+1	-1	+1	11.18	0.011633
ab	+1	+1	-1	+1	-1	-1	-1	12.42	0.002033
c	-1	-1	+1	+1	-1	-1	+1	7.85	0.062025
ac	+1	-1	+1	-1	+1	-1	-1	9.19	0.028292
bc	-1	+1	+1	-1	-1	+1	-1	14.76	0.009200
abc	+1	+1	+1	+1	+1	+1	+1	16.64	0.019667
Contrast	6.12	19.40	6.28	0.12	0.32	9.32	0.96	90.60	0.173634

A test statistic for $H_0: \Psi = 0$ versus $H_a: \Psi \neq 0$ is $T = C/\sqrt{32(MSE)}$, where T has a t distribution with 24 degrees of freedom. The p-value is $2P(T \geq |t|)$. The first of the 7 tests follows.

$H_{01}: \Psi_A = 0; H_{a1}: \Psi_A \neq 0$

Test statistic: $T = C_A/\sqrt{32(MSE)}$; $\nu = 24$

Analysis: From the preceding table, $C_A = 6.12$ and the observed value of MSE is $0.173634/8 \approx 0.021704$. So, $t \approx 6.12/\sqrt{(32)(0.021704)}$ $= 6.12/\sqrt{0.694528} \approx 7.3436$. The p-value is $2P(T > 7.3436) \approx 0.0000$. We reject H_{01} and conclude that A is an active contributor to the variability in the trench height.

For each of the 7 hypothesis tests, the observed value of T and the p-value of the test are summarized in the following table. A, B, C, and BC are significant. Thus, we should analyze the means at the 2 levels of A and the means at the 4 levels of BC.

Effect	t	p-value	Effect	t	p-value
A	7.3436	0.0000	AC	0.3840	0.7044
B	23.2786	0.0000	BC	11.1833	0.0000
AB	0.1440	0.8867	ABC	1.1519	0.2607
C	7.5355	0.0000			

Beginning with the profile plot of the sample means for power (*A*) given to the right, and noting that the trench height is to be 3 microns, we recommend use of the high power level.

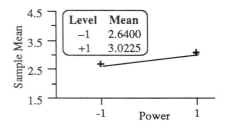

Level	Mean
−1	2.6400
+1	3.0225

The following profile plot for the sample means associated with the interaction between the 2 flow rates (*BC*) seems to indicate that the low oxygen flow rate and the high SF6 flow rate should be used. Before making any recommendations, we will see if Tukey's procedure supports such claims.

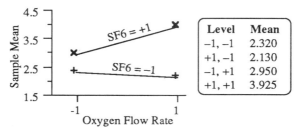

Level	Mean
−1, −1	2.320
+1, −1	2.130
−1, +1	2.950
+1, +1	3.925

Suppose the multiple comparison of the $k = 4$ means is to be conducted at a 5% multiple comparison rate. Since 24 degrees of freedom are associated with *MSE*, let $\nu = 24$. Using Appendix S, the values of $Q_{0.95}$ at $\nu = 20$ and $\nu = 40$ are 3.96 and 3.79, respectively. Linear interpolation gives $Q_{0.95} \approx 3.93$ at $\nu = 24$. Since $n = 8$ observations were obtained at each of the 4 treatment combinations and $MSE \approx 0.021704$, $w \approx 3.93\sqrt{0.021704/8} \approx 0.2047$. If two sample means differ by at least 0.2047 microns, conclude that the population mean associated with the larger sample mean is the greater of the two population means.

The observed means for the treatment combinations 1 = (−1, −1), 2 = (+1, −1), 3 = (−1, +1), and 4 = (+1, +1) are 2.320, 2.130, 2.950, and 3.925, respectively. [**Note:** The pairs are of the form (*C, B*) with *C* the oxygen flow rate and *B* the SF6 flow rate.] Placing these means in ascending order and comparing the differences to 0.2047 gives the following graphic for Tukey's procedure.

Population	Label	and	Sample	Mean
2 = (+1, −1)	1 = (−1, −1)		3 = (−1, +1)	4 = (+1, +1)
2.13	2.32		2.95	3.925

With a 5% experimentwise error, $\mu_1 < \mu_3$, $\mu_1 < \mu_4$, $\mu_2 < \mu_3$, $\mu_2 < \mu_4$, and $\mu_3 < \mu_4$. There is insufficient evidence to declare μ_1 and μ_2

significantly different. Since μ_3 is significantly different from the other means and $\bar{y}_3$ is nearest 3 microns, the high SF6 flow rate and the low oxygen flow rate are recommended.

In summary, using A (power) and B (SF6 flow rate) at their high settings with C (oxygen flow rate) at the low setting appears to result in an average height close to 3.0 microns.

(b) The following JMP summary is obtained for

$$Y = \beta_0 + \beta_1 x_1 + \beta_2 x_2 + \beta_3 x_1 x_2 + \beta_4 x_3 + \beta_5 x_1 x_3 + \beta_6 x_2 x_3 + \beta_7 x_1 x_2 x_3 + \varepsilon$$

with

$$x_1 = \begin{cases} -1 \text{ at the low power level} \\ +1 \text{ at the high power level} \end{cases},$$

$$x_2 = \begin{cases} -1 \text{ at the low SF6 flow rate} \\ +1 \text{ at the high SF6 flow rate} \end{cases}, \text{ and}$$

$$x_3 = \begin{cases} -1 \text{ at the low oxygen flow rate} \\ +1 \text{ at the high oxygen flow rate} \end{cases}.$$

Power, both flow rates, and the interaction between the 2 flow rates are significant at any reasonable value of α. [**Note:** The t ratios and p-values are the same as those obtained in (a).] Thus, we should analyze the means at the 2 levels of A (power) and the means at the 4 levels of BC (interaction between the flow rates).

Term	Estimate	Std Error	t Ratio	Prob>|t|
Intercept	2.83125	0.026043	108.7130	<.0001
Power	0.19125	0.026043	7.3435	<.0001
SF6_Flow	0.60625	0.026043	23.2785	<.0001
Power*SF6	0.00375	0.026043	0.1440	0.8867
Oxy_Flow	0.19625	0.026043	7.5355	<.0001
Power*Oxy	0.01000	0.026043	0.3840	0.7044
SF6*Oxy	0.29125	0.026043	11.1833	<.0001
Power*SF6*Oxy	0.03000	0.026043	1.1519	0.2607

Analyses of A and BC were presented in (a). From those results, it appears that an average height close to 3.0 microns will result when the integrated circuits are processed using A (power) and B (SF6 flow rate) at their high settings with C (oxygen flow rate) at the low setting. [**Note:** From the preceding summary,

$$\hat{y} \approx 2.83125 + 0.19125 x_1 + 0.60625 x_2 + 0.00375 x_1 x_2$$
$$+ 0.19625 x_3 + 0.01000 x_1 x_3 + 0.29125 x_2 x_3 + 0.03000 x_1 x_2 x_3.$$

At $(+1, +1, -1)$, $x_1 = 1$, $x_2 = 1$, $x_3 = -1$, and

$$\hat{y} \approx 2.83125 + 0.19125(1) + 0.60625(1) + 0.00375(1)(1)$$
$$+ 0.19625(-1) + 0.01000(1)(-1) + 0.29125(1)(-1)$$
$$+ 0.03000(1)(1)(-1) = 3.10500.]$$

189

A coefficient of determination of $R^2 = 0.970118$ is given in the following JMP summary of fit. Power, SF6 flow rate, oxygen flow rate, and the 4 interactions account for approximately 97% of the variability in the trench height readings. The groups of points in the following scatter plot of height versus predicted height form a linear pattern along the diagonal line $y = \hat{y}$. The 3-factor model with interactions provides a reasonable fit to the data.

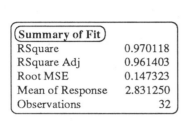

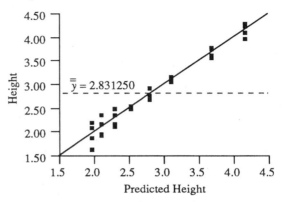

The following box plot, normal quantile plot, and Shapiro-Wilk W test were obtained using JMP Version 3.1.5. The normal quantile plot includes the Lillefors graph as a test for normality. If any points fall outside the Lillefors bounds (indicated by the dotted curves), the sampled population should be considered nonnormal. Otherwise, the sample could have come from a normally distributed population. In this case, the plot supports an assumption of normality. The Shapiro-Wilk W test (p-value = 0.8716) also supports that assumption. The box plot is reasonably symmetric and reveals no outliers. The assumption of normality should be adequately satisfied.

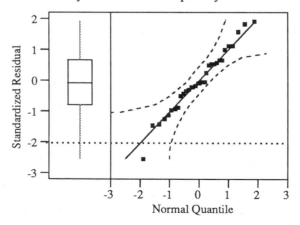

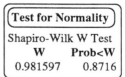

The following plot of the standardized residuals versus the predicted heights reveals one standardized value below −2 and seems to indicate more variability in trench height when the predicted height is near 2 microns. Those predicted heights occur when power and SF6 flow are at their low levels. The small standardized residual (−2.52771) is associated with the height 1.64 obtained at the point (−1, −1, +1). A study of the conditions at the times the integrated circuit having that trench height was processed and the measurement was obtained should be thoroughly investigated.

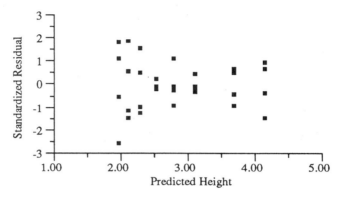

Not only does the preceding scatter plot provide some evidence of unequal variances, but the sample variances summarized in (a) are quite variable. To investigate further, we consider the ranges of the 8 samples of 4 standardized residuals. Each of those ranges is within the control limits given in the following JMP range chart. Thus, the assumption of equal variances should be adequately satisfied.

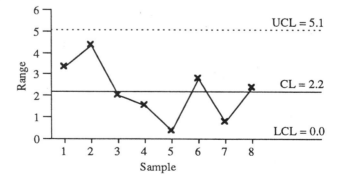

13.13 **(a)** Multiplying corresponding elements in the A, B, C, and D columns of the following table, a column of +1's is obtained. No informa-tion about the low level of $ABCD$ is available, indicating that $ABCD$ is confounded with blocks.

tc	Alternate Description of tc				Contrast Coefficients			Total
	A	B	C	D	•	•	$ABCD$	
(1)	−1	−1	−1	−1	•	•	+1	6.08
ab	+1	+1	−1	−1	•	•	+1	6.43
ac	+1	−1	+1	−1	•	•	+1	6.09
bc	−1	+1	+1	−1	•	•	+1	6.12
ad	+1	−1	−1	+1	•	•	+1	6.68
bd	−1	+1	−1	+1	•	•	+1	6.73
cd	−1	−1	+1	+1	•	•	+1	6.77
abcd	+1	+1	+1	+1	•	•	+1	6.23

(b)

$$
\begin{array}{llll}
I & = ABCD & & \\
A & = A^2BCD & = BCD & \\
B & = AB^2CD & = ACD & \\
AB & = A^2B^2CD & = CD &
\end{array}
\qquad
\begin{array}{llll}
C & = ABC^2D & = ABD \\
AC & = A^2BC^2D & = BD \\
BC & = AB^2C^2D & = AD \\
D & = ABCD^2 & = ABC
\end{array}
$$

(c)

Trt. Cmb.	Linear Contrast Coefficients							Total
	A	B	C	D	AB	AC	AD	
(1)	−1	−1	−1	−1	+1	+1	+1	6.08
ab	+1	+1	−1	−1·	+1	−1	−1	6.43
ac	+1	−1	+1	−1	−1	+1	−1	6.09
bc	−1	+1	+1	−1	−1	−1	+1	6.12
ad	+1	−1	−1	+1	−1	−1	+1	6.68
bd	−1	+1	−1	+1	−1	+1	−1	6.73
cd	−1	−1	+1	+1	+1	−1	−1	6.77
abcd	+1	+1	+1	+1	+1	+1	+1	6.23
Est. Eff.	−0.27	−0.11	−0.71	1.69	−0.11	−0.87	−0.91	51.13
Alias	BCD	ACD	ABD	ABC	CD	BD	BC	

Since only one observation was obtained per treatment combination, the estimated effect is the corresponding contrast in the totals. For example, $c_{AD} = 6.08 - 6.43 - 6.09 + 6.12 + 6.68 - 6.73 - 6.77 + 6.23 = -0.91$ is the estimated effect of AD.

(d)

i	$y_{(i)}$	p_i	z_i	$(z_i, y_{(i)})$
1	−0.91	0.125	−1.15	(−1.15, −0.91)
2	−0.87	0.250	−0.67	(−0.67, −0.87)
3	−0.71	0.375	−0.32	(−0.32, −0.71)
4	−0.27	0.500	0.00	(0.00, −0.27)
5	−0.11	0.625	0.32	(+0.32, −0.11)
6	−0.11	0.750	0.67	(+0.67, −0.11)
7	1.69	0.875	1.15	(+1.15, +1.69)

The points in the following normal quantile plot are summarized in the preceding table (with values of z rounded to the nearest 100th). The effect of D (which is aliased with ABC) lies well off a line through the main body of points. The sample means associated with D are $(6.08 + 6.43 + 6.09 + 6.12)/4 = 6.18$ at the low level and $(6.68 + 6.73 + 6.77 + 6.23)/4 = 6.6025$ at the high level. If the effect of ABC can be assumed insignificant or zero, and if a larger (true) average response is better, the higher level of D should be used.

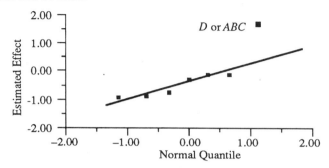

13.19 (a) The aliases of an effect are obtained by multiplying each element of the defining relationship by that effect. For example, the aliases of AD are BC, $ABEF$, and $CDEF$ because $(AD)(I) = AD$, $(AD)(ABCD) = A^2BCD^2 = BC$, $(AD)(BDEF) = ABD^2EF = ABEF$, and $(AD)(ACEF) = A^2CDEF = CDEF$. That is, $AD = BC = ABEF = CDEF$. Using this procedure with the other effects produces the following summary of the alias structure.

I	$= ABCD$	$= BDEF$	$= ACEF$	AC	$= BD$	$= ABCDEF$	$= EF$
A	$= BCD$	$= ABDEF$	$= CEF$	AD	$= BC$	$= ABEF$	$= CDEF$
B	$= ACD$	$= DEF$	$= ABCEF$	AE	$= BCDE$	$= ABDF$	$= CF$
C	$= ABD$	$= BCDEF$	$= AEF$	BE	$= ACDE$	$= DF$	$= ABCF$
D	$= ABC$	$= BEF$	$= ACDEF$	BF	$= ACDF$	$= DE$	$= ABCE$
E	$= ABCDE$	$= BDF$	$= ACF$	CE	$= ABDE$	$= BCDF$	$= AF$
F	$= ABCDF$	$= BDE$	$= ACE$	ABE	$= CDE$	$= ADF$	$= BCF$
AB	$= CD$	$= ADEF$	$= BCEF$	ADE	$= BCE$	$= ABF$	$= CDF$

(b) The defining contrast partitions the 64 treatment combinations into the following four blocks of 16 treatment combinations each.

Block 1: all treatment combinations that share an even number of letters with $ABCD$ and an even number of letters with $BDEF$.

Block 2: all treatment combinations that share an even number of letters with $ABCD$ and an odd number of letters with $BDEF$.

Block 3: all treatment combinations that share an odd number of letters with $ABCD$ and an even number of letters with $BDEF$.

Linear Contrast Coefficient

tc	A	B	C	D	E	F	AB	AC	AD	AE	BE	BF	CE	ABE	ADE	Total
(1)	-1	-1	-1	-1	-1	-1	+1	+1	+1	+1	+1	+1	+1	-1	-1	T_1
ac	+1	-1	+1	-1	-1	-1	-1	+1	-1	-1	+1	+1	-1	+1	+1	T_2
bd	-1	+1	-1	+1	-1	-1	-1	+1	-1	+1	-1	-1	+1	+1	+1	T_3
abcd	+1	+1	+1	+1	-1	-1	+1	+1	+1	-1	-1	-1	-1	-1	-1	T_4
abe	+1	+1	-1	-1	+1	-1	+1	-1	-1	+1	+1	-1	-1	+1	-1	T_5
bce	-1	+1	+1	-1	+1	-1	-1	-1	+1	-1	+1	-1	+1	-1	+1	T_6
ade	+1	-1	-1	+1	+1	-1	-1	-1	+1	+1	-1	+1	-1	-1	+1	T_7
cde	-1	-1	+1	+1	+1	-1	+1	-1	-1	-1	-1	+1	+1	+1	-1	T_8
abf	+1	+1	-1	-1	-1	+1	+1	-1	-1	+1	-1	+1	+1	-1	+1	T_9
bcf	-1	+1	+1	-1	-1	+1	-1	-1	+1	+1	-1	+1	-1	+1	-1	T_{10}
adf	+1	-1	-1	+1	-1	+1	-1	-1	+1	-1	+1	-1	+1	+1	-1	T_{11}
cdf	-1	-1	+1	+1	-1	+1	+1	-1	-1	+1	+1	-1	-1	-1	+1	T_{12}
ef	-1	-1	-1	-1	+1	+1	+1	+1	+1	-1	-1	-1	-1	+1	+1	T_{13}
acef	+1	-1	+1	-1	+1	+1	-1	+1	-1	+1	-1	-1	+1	-1	-1	T_{14}
bdef	-1	+1	-1	+1	+1	+1	-1	+1	-1	-1	+1	+1	-1	-1	-1	T_{15}
abcdef	+1	+1	+1	+1	+1	+1	+1	+1	+1	+1	+1	+1	+1	+1	+1	T_{16}
Alias	BCD CEF ABDEF	ACD DEF ABCEF	ABD AEF BCDEF	ABC BEF ACDEF	ACF BDF ABCDE	ACE BDE ABCDF	CD BCEF ADEF	BD EF ABCDEF	BC ABEF CDEF	CF BCDE ABDF	DF ACDE ABCF	DE ABCE ACDF	AF ABDE BCDF	CDE BCF ADF	BCE ABF CDF	

Orthogonal Table for Problem 13.19(c)

Block 4: all treatment combinations that share an odd number of letters with *ABCD* and an odd number of letters with *BDEF*.

Since (1) shares an even number of letters (i.e., 0) with both *ABCD* and *BDEF*, a listing of the elements in Block 1 has been requested. Comparing each of (1), *a, b, ab, c, ac, bc, abc, d, ad, bd, abd, cd, acd, bcd, abcd, e, ae, be, abe, ce, ace, bce, abce, de, ade, bde, abde, cde, acde, bcde, abcde, f, af, bf, abf, cf, acf, bcf, abcf, df, adf, bdf, abdf, cdf, acdf, bcdf, abcdf, ef, aef, bef, abef, cef, acef, bcef, abcef, def, adef, bdef, abdef, cdef, acdef, bcdef,* and *abcdef* with *ABCD* and *BDEF*, that block is {(1), *ac, bd, abcd, abe, bce, ade, cde, abf, bcf, adf, cdf, ef, acef, bdef, abcdef*}. This block is called the *principal block*.

(c) The orthogonal table for the one-fourth replicate in (b) is on page 194. Entries in the interaction columns were obtained by multiplying corresponding row entries in the associated main effects columns. For example the entries in row *abf* of columns *A*, *B*, and *E* are +1, +1, and −1, respectively. Thus, the entry in row *abf* of column *ABE* is (+1)(+1)(−1) = −1.

(d) Multiplying corresponding entries in the *A*, *B*, *C*, and *D* columns of the orthogonal table in (c) produces a column of +1's. No information about the low level of *ABCD* is available, indicating that *ABCD* is confounded with blocks. The same is true of the product of corresponding entries in the individual effect columns associated with *BDEF* and *ACEF*. Thus, *BDEF* and *ACEF* are also confounded with blocks.

13.21 An orthogonal table for the combined half-replicates (blocks) is given on page 196, which follows. Information associated with the half-replicate considered in Problem 13.20 is in the shaded region of that table. Notice that the 8 entries in the *ABCD* column are +1 for the half-replicate in Example 13.10 and −1 for the half-replicate in Problem 13.20. Thus, any information about *ABCD* is confounded with information about the two blocks.

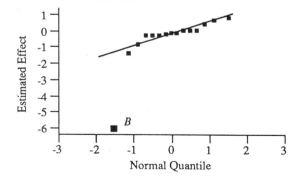

Linear Contrast Coefficient

t_c	A	B	C	D	AB	AC	AD	BC	BD	CD	ABC	ABD	ACD	BCD	ABCD	Total
(1)	-1	-1	-1	-1	+1	+1	+1	+1	+1	+1	-1	-1	-1	-1	+1	1.49
ab	+1	+1	-1	-1	+1	-1	-1	-1	-1	+1	-1	-1	+1	+1	+1	0.63
ac	+1	-1	+1	-1	-1	+1	-1	-1	+1	-1	-1	+1	-1	+1	+1	1.40
ad	+1	-1	-1	+1	-1	-1	+1	+1	-1	-1	+1	-1	-1	+1	+1	1.09
bc	-1	+1	+1	-1	-1	-1	+1	+1	-1	-1	-1	+1	+1	-1	+1	0.69
bd	-1	+1	-1	+1	-1	+1	-1	-1	+1	-1	+1	-1	+1	-1	+1	0.69
cd	-1	-1	+1	+1	+1	-1	-1	-1	-1	+1	+1	+1	-1	-1	+1	1.60
abcd	+1	+1	+1	+1	+1	+1	+1	+1	+1	+1	+1	+1	+1	+1	+1	0.64
a	+1	-1	-1	-1	-1	-1	-1	+1	+1	+1	+1	+1	+1	-1	-1	1.18
b	-1	+1	-1	-1	-1	+1	+1	-1	-1	+1	+1	+1	-1	+1	-1	0.74
c	-1	-1	+1	-1	+1	-1	+1	-1	+1	-1	+1	-1	+1	+1	-1	1.60
d	-1	-1	-1	+1	+1	+1	-1	+1	-1	-1	-1	+1	+1	+1	-1	1.50
abc	+1	+1	+1	-1	+1	+1	-1	+1	-1	-1	+1	-1	-1	-1	-1	0.69
abd	+1	+1	-1	+1	+1	-1	+1	-1	+1	-1	-1	+1	-1	-1	-1	0.62
acd	+1	-1	+1	+1	-1	+1	+1	-1	-1	+1	-1	-1	+1	-1	-1	1.43
bcd	-1	+1	+1	+1	-1	-1	-1	+1	+1	+1	-1	-1	-1	+1	-1	0.65
Est. Effect	-1.28	-5.94	0.76	-0.20	0.90	0.52	-0.04	-0.78	-0.10	0.08	-0.18	0.10	0.08	-0.14	-0.18	16.64

Orthogonal Table for Problem 13.21

A normal quantile plot of the effects summarized on page 196 is presented on page 195. The point associated with the effect of B falls well off a line through the main body of points. We conclude that hold pressure (B) is more influential than the other effects.

For the sample data, the average shrinkage is

$(1.49 + 1.40 + 1.09 + 1.60 + 1.18 + 1.60 + 1.50 + 1.43)/8 = 11.29/8 = 1.41125$

at the low hold pressure and

$$(16.64 - 11.29)/8 = 5.35/8 = 0.66875$$

at the high hold pressure. Since lesser shrinkage is preferred, the high hold pressure should be used.

Section 13.3

13.25

Trt. Cmb.	Linear Contrast Coefficients							Response
	A	B	C	D	E	F	G	
(1)	−1	−1	−1	−1	−1	−1	−1	16
defg	−1	−1	−1	+1	+1	+1	+1	17
bcfg	−1	+1	+1	−1	−1	+1	+1	12
bcde	−1	+1	+1	+1	+1	−1	−1	6
aceg	+1	−1	+1	−1	+1	−1	+1	6
acdf	+1	−1	+1	+1	−1	+1	−1	68
abef	+1	+1	−1	−1	+1	+1	−1	42
abdg	+1	+1	−1	+1	−1	−1	+1	26
Est. Eff.	91	−21	−9	41	−51	85	−71	193

The normal quantile plot of the estimated effects summarized in the preceding table seems to indicate that A, D, and F are active contributors to the variability in the number of defective tiles per 100.

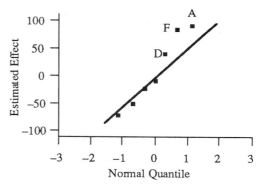

Each trial resulted in 100 tiles, so 400 tiles were associated with each level of a factor. For A, 51 defective tiles were obtained at the low level and 142 defective tiles were obtained at the high level. At the low and high levels of D, the numbers of defective tiles were 76 and 117, respectively. Also, 132 defective tiles were obtained at the low level of G and 61 were obtained at the high level. Future processing with A low, D low, and G high should be considered.

13.27

Source	df	SS	MS	F-ratio	p-value
Thickness	1	109,203.33	109,203.33	140.44	0.0000
Paint	2	193,032.47	96,516.24	124.12	0.0000
Interaction	2	89,302.07	44,651.04	57.42	0.0000
Error	24	18,662.00	777.58		
Total	29	410,199.87			

The interaction effect is significant (p-value = 0.0000), so a comparison of the 6 means should follow. The following profile plot of the means seems to indicate that paint 1 could be used at the 16 micron thickness.

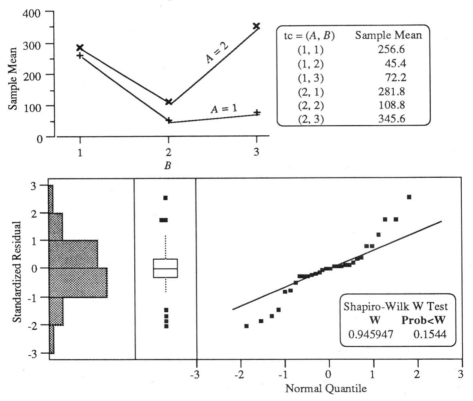

The preceding box plot of the standardized residuals reveals several outliers. The associated readings should be investigated before making a final decision.

The Shapiro-Wilk W test (p-value ≈ 0.15) provides no strong evidence for rejection of normality. However, the pattern in the normal quantile plot seems to indicate that the normality should not be assumed. Combining this information with that from the box plot, we will not assume normality.

Since a normality assumption seems unreasonable, a multiple comparison of the means using Tukey's procedure will not be conducted. We do, however, have reason to believe that paint 1 could be used at the 16 micron thickness. Follow-up experiments should be conducted using paint 1 at that thickness.

Section 13.5

13.29

Source	df	SS	MS	F-ratio	p-value
Stress (T)	3	30,316.750	10,105.583	3.7256	0.0332
Current (C)	1	28,203.125	28,203.125	10.3977	0.0053
T*C	3	27,826.125	9,275.375	3.4196	0.0429
Switching (W)	1	79,600.500	79,600.500	29.3465	0.0001
T*W	3	18,011.750	6,003.917	2.2135	0.1261
C*W	1	48,205.125	48,205.125	17.7719	0.0007
T*C*W	3	5,111.125	1,703.708	0.6281	0.6073
Error	16	43,399.000	2,712.438		
Total	31	280,673.500			

For $\alpha = 0.05$, the effect of the interaction between stress and current is statistically significant, as is the effect of the interaction between current and switching.

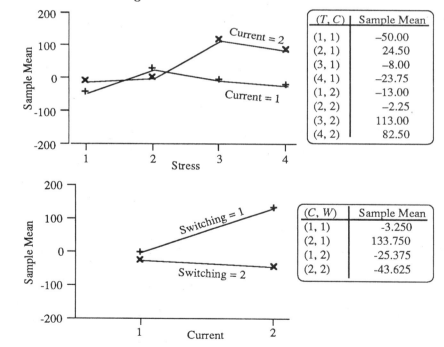

The preceding profile plot of current versus environmental stress seems to indicate that the average decrease in output current is greater when the high current level is used at the high levels of environmental stress. The other profile plot seems to indicate that use of the lower switching level with the higher current level produces a greater average decrease in output current.

The following graphics and Shapiro-Wilk W test were obtained using JMP Version 3.1.5. The normal quantile plot includes the Lillefors graph as a test for normality. If any points fall outside the Lillefors bounds, the sampled population should be considered nonnormal. In this case, the plot supports a normality assumption, as do the histogram, box plot, and Shapiro-Wilk W test (p-value = 0.7910). Since a normality assumption seems reasonable, we can use Tukey's multiple comparison procedure with the pairs of means.

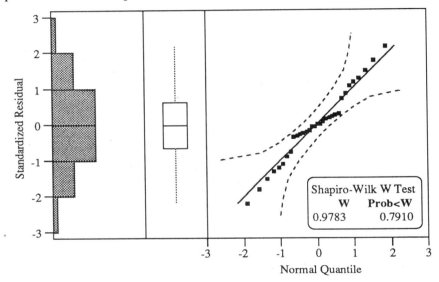

To compare the 8 means associated with pairs of the form (stress, current), we first note that the error sum of squares has 16 degrees of freedom and $MSE = 2712.438$. Using Appendix S with $k = 8$, $v = 16$, and $\alpha = 0.05$, we find $Q_{0.95} = 4.90$. Since $MSE = 2712.438$ and $n = 4$ observations were obtained at each of the 8 treatment combinations, $w = Q_{0.95}\sqrt{MSE/n} = 4.90\sqrt{2712.438/4} \approx 127.599$. Placing the 8 sample means (see the first of the preceding profile plots) in ascending order and comparing the differences to 127.599 gives the following graphic for Tukey's multiple comparison procedure. With a 5% experimentwise error, we conclude that $\mu_7 > \mu_1$, $\mu_8 > \mu_2$, and $\mu_8 > \mu_1$. The average decrease in output current when stress is at level 3 or 4 and input current is high exceeds that average when stress is at the low level and input current is low. Also, the average decrease in output current when stress is at level 3 and input current is high exceeds that average when stress is at level 4 and input current is low.

200

Population Label and Sample Mean							
1 = (1, 1)	2 = (4, 1)	3 = (1, 2)	4 = (3, 1)	5 = (2, 2)	6 = (2, 1)	7 = (4, 2)	8 = (3, 2)
−50.00	−23.75	−13.00	−8.00	−2.25	24.50	82.50	113.00

Now consider the 4 means associated with pairs of the form (current, switching). Using Appendix S with $k = 4$, $v = 16$, and $\alpha = 0.05$, we find $Q_{0.95} = 4.05$. Since $n = 8$ observations were obtained at each of the 4 treatment combinations,

$$w = Q_{0.95}\sqrt{MSE/n} = 4.05\sqrt{2712.438/8} \approx 74.574.$$

Placing the 4 sample means (see the second of the preceding profile plots) in ascending order and comparing the differences to 74.574 gives the following graphic for Tukey's multiple comparison procedure. With a 5% experiment-wise error, we conclude that the average decrease in output current is greatest when the input current is high and switching is low.

Population Label and Sample Mean			
1 = (2, 2)	2 = (1, 2)	3 = (1, 1)	4 = (2, 1)
−43.625	−25.375	−3.250	133.750

201